The Art of Coding

The Language of Drawing, Graphics, and Animation

The Art of Coding

The Language of Drawing, Graphics, and Animation

Mohammad Majid al-Rifaie
Anna Ursyn
Theodor Wyeld

CRC Press is an imprint of the
Taylor & Francis Group, an informa business
A CHAPMAN & HALL BOOK

First edition published 2020
by CRC Press
6000 Broken Sound Parkway NW, Suite 300, Boca Raton, FL 33487-2742

and by CRC Press
2 Park Square, Milton Park, Abingdon, Oxon OX14 4RN

ISBN: 978-0-367-90037-3 (hbk)
ISBN: 978-1-138-62964-6 (pbk)
ISBN: 978-1-315-21033-9 (ebk)

To our students

Contents

List of Codes

Foreword

Since I can remember, I was always saying that the "code is art"! Coding activates creative thinking, and as Steve Jobs once said: "*Everybody in this country should learn to program a computer, because it teaches you how to think*" – and I fully agree with that. You can learn how to use coding for drawing figures, creating graphics layouts, or going further and assembling interactive animations – then, if you know how to draw a pixel, you can draw anything that your creative imagination brings to the table.

In this sense, the book that you have in your hands, *The Art of Coding: The Language of Drawing, Graphics, and Animation*, is a very special one. It consists of lots of complete code examples that will act as a guiding voice, showing you what is possible using creative thinking – a pure inspiration. It will teach you how to code animations, create audio visual installations, build 3D visualisers, morphing create weather tree applications, connect and interact with embedded electronic systems such as Arduino, or execute various image processing routines, and, much, much more. This book will allow you to experiment using guided practice with different programming languages and data structures such as text, images, animations, use of APIs, sound data, and music. It will open new horizons and an understanding that coding can be used in various imaginative and inspirational ways, and that it is accessible to creative people, and indeed everyone, not only software engineers or computer science professionals. I strongly encourage you to experiment with various code examples; even if you break them, in the end, you learn by practice.

This book is written in very positive and inspiring ways, and educates not only on technological aspects, but also historical facts, art, science, and engineering. It is written by three experts in the field – educators, artists, and computer scientists with diverse knowledge and connected across continents. It is highly recommended to everyone: Artists, scientists, designers, engineers, educators, managers – whoever you are, you will get inspired by possibilities of coding after reading this book.

Tomasz Bednarz

Director and Head of Visualisation at the Expanded Perception and Interaction Centre (EPICentre) at the UNSW Art & Design | Visual Analytics Team Leader at the CSIRO Data61 | ACM SIGGRAPH Asia 2019 Conference Chair | Demoscene Coder | Artist-Scientist.

Preface

THERE are countless new apps, methods-related solutions, and web solutions that create opportunities for personal, technological, and materials-related growth, especially when programming for new startups. This book is aimed at making programming easier by presenting it in a visual way. The authors' intention is to help the reader understand the core programming concepts for art, web, and everyday applications including pervasive, ubiquitous, and wearable apps. We also discuss how applying pictures, analogies, and metaphors that refer to familiar sensory faculties helps. Making the programming processes visual helps the reader understand the underlying concepts, and gain the knowledge needed, faster and easier. This book also addresses readings in art, science, mathematics, and other subjects, as well as some broader issues such as multicultural, environmental, and social objectives.

Being able to program requires abstract thinking, math skills, spatial ability, logical thinking, imagination, and creativity. All these abilities can be acquired with practice. They can be mastered by practical exposure to art, music, and literature. In this book, we discuss art, poetry, and other forms of writing, while pondering difficult concepts in programming. The book discusses how we use our senses in the process of learning computing and programming, as well as how visual and verbal ways of studying these processes can involve artistic creation. Programming can be seen as a set of concepts linked by a common framework. The same image can be created with different computer languages. Each time a different compiler might be used with a different environment set up – with different concepts between each computing language, its origin, which group of languages it belongs to, its strengths and weaknesses, and so on. To make this easier, we have focused on visualisation of the processes as an outcome that serves various creative goals or applications.

Students in programming classes often participate in exercises that prepare them to master a computer language but not to grasp the aesthetic concepts, composition of the project, or its conceptual meaning. For example, students might be asked to perfect their computer science skills by writing a program for a colourful ball that is bouncing in a closed rectangular area and changing its colour after each contact with a border. Students produce similar, easy to assess solutions; they may work on the changing reflections of a background on the balls, but they are usually not expected to create anything original or new. The task for the programmer can be solved with various coding strategies.

Many times individual programmers may come to the same solution using different programs. The final products look the same while the code itself may vary. Examining the ways various students solve the same assignment might be one of the best practices in learning how to code. Contrary to these practices, when confronted with a learning project to be depicted with the use of programming, students employ their cognitive thinking abilities while they strengthen their abstract thinking and inventive imaging ability; they write unique solutions in their programs to create one-of-a-kind computer graphics.

Visual communication goes beyond verbal, and verbal beyond visual. There are several reasons for combining the forces of the artist, designer, and programmer. First of all, communication has become more and more visual, especially on the internet, on the phone, and other portable devices. Traditional divisions, such as Mac–PC, are lessening in importance because programmers need to understand and apply a visual language while designers and artists need to apply technology. Both need to understand coding and stay current with the opportunities offered by computers. But comprehension of a code does not always go together with attentiveness to visual characteristics of the programmed products. This book is aimed at directing readers' awareness to a visual way of thinking. Along with a presentation of basic facts and theories, it presents a series of learning projects that focus on organising concepts and data into code and creating practical applications for technologically-based creativity.

Multisensory-based perception is conducive to learning. We can see a tendency to take more senses into consideration when designing projects and apps. This tendency may be seen in many 3D web, augmented reality (AR) and virtual reality (VR) projects, holographic productions, fabrication, and other transitions to physical computing. Within this paradigm, the Apple Computer Company has developed connectors that cannot be misused: Only the right connection works. Also icons, symbols, and signs are internationally recognisable in a fast and efficient way within this set of concepts. However, symbols are different in specific fields: A letter C means quite another thing in music, in mathematics, in chemistry, or in C, C++, or C# coding.

Projects and products' aesthetics should be linked with usability. This may be applied to the web (the way of delivery), marketing (product promotion), product design (their functioning), and apps (which relates to all of the above). In other words, an app should not need any explanation about how it works – it should be largely intuitive. For these reasons, coding literacy is needed by everyone. One may say creativity plus coding literacy results in a new kind of artistic creation. Indeed, many want to become artists who push art to a new realm by involving coding and technology. This means it is necessary to learn how to code, and use supporting software applications with coding. Artworks accepted at some competitive shows and galleries (for example, Ars

Electronica) are mostly time-based and interactive, while even stunning static images are often no longer enough.

The key features that describe this book can be listed as follows:

1. Coding is presented and taught in a visual way.
2. Instruction is focused on the elegance behind coding and the outcome.
3. Stress is put on many types of outcomes and options for coding.
4. Sample codes are provided for the learners, along with the outcomes.
5. Instruction is aimed at strengthening students' abstract thinking and creativity.

Functional requirements for coding stimulate technical development. While coding techniques change, they can assume new meanings, scope, purpose, and role. First of all, programming becomes interdisciplinary when it serves projects co-developed by specialists in different disciplines; it may link the ways of thinking characteristic of these disciplines, methods, and trends – characteristic of the individual disciplines for which it is serving. Data can be collected through collaboration; however, distinct shortcuts adopted by various specialists may cause problems in understanding their code. Moreover, particular programmers may achieve the same solution by writing different code. Hence, discussion of a short, elegant program as contrasted with another robust one may be an interesting way of supporting the students' learning process while teaching to code.

This book is about possible choices in creating by coding some two-dimensional shapes and solid 3D forms that are changeable over time or by interactivity. This book contains a collection of learning projects for the students, instructors, and teachers to select specific themes from. Problems and projects are aimed at making the learning process entertaining while also involving social exchange and sharing. An accompanying website hosts the code and outcomes available to the readers. Learning projects may be useful for people working in various disciplines and jobs. For this reason, a selected image may be presented, or reworked in different programs: For example, an image of a person can be shown as an adult, a warrior, or a mannequin, and then presented in different media. A large part of the learning projects in this book is focused on creating natural shapes and forms: For example, a form of a mushroom might inspire both the builders of ancient Ionian columns and also designers of umbrellas.

Text and speech involve different senses and also cortex areas in the brain; therefore, the cognitive processes may be different. We use text to create communication between a person and a machine. If we write a poem about a white horse, a reader will visualise a white horse. However, the image of a horse would be different depending on one's imagination and the context of this horse. If one says, "this is a white horse," the imagined horse would look different depending on the timbre of the voice, the look, appearance, expression, or mien of the speaker. If it is written, the literary style could create different features both of a horse and its surroundings. When we read a book first and then see the movie, the characters' representations in the film do not always fit our imagined idea of those individuals. Similarly, if we write a program for a computer, it will generate an image of a white horse; the differences (variations) in the code or in the computer language used will result in the differing printout appearances.

The goal of this book is to facilitate the learning process of coding; it uses visualisations, aspires to inspire, and pushes coding beyond a technical task by treating the process as a creative approach to problem solving.

This book can be used as a resource book, a supplemental book for teachers, developers, designers, and for people interested in computer science, computer graphics, digital media, or interdisciplinary studies. People from production companies can see how coding supports software, and how it generates inspiration for new solutions. A possible application of this material pertains to journalism, nursing, art, architecture, business, and computer science, among many other disciplines. It may serve for media courses across departments, for hobbyists, and middle and high school students. We hope that coding will soon enter the curricula starting from kindergarten and elementary schools. When used as a required textbook, this book can serve courses such as CS 101, digital media, computer graphics, visual design, CIS, instructional technology, communication studies, and art classes.

Contributors

Erik Brunwald
University of Utah
USA

Alireza Ebrahimi
State University of New York Old Westbury
SUNY, USA

Md Fahimul Islam
Walgreens
Illinois, USA

Stuart Smith
University of Massachusetts Lowell
USA

Scott Peckover
Flinders University
Australia

Mohammad Ali Javaheri Javid
Goldsmiths, University of London
London, UK

Thao Luo
Flinders University
Australia

Wilson Bolas Tolentino da Silva
Goldsmiths, University of London
London, UK

Ahmed Aber
The University of Sheffield
Sheffield, UK

SPECIAL THANKS TO:

Zameer Faiz
Freelancer.com
Australia

Gurtaj Singh
Freelancer.com
Australia

Platon Pronko
blog.rogach.org
Czech Republic

Brandon Mckay
Flinders University
Australia

AND OUR STUDENTS:

Matt Anderson
University of Northern Colorado

Amanda Betts
University of Northern Colorado

Christian Eggers
University of Northern Colorado

Sean Flannery
University of Northern Colorado

Moises Gomez
University of Northern Colorado

Kyle Hathcoat
University of Northern Colorado

Morgan Hurtado
University of Northern Colorado

Jenny Lee
University of Northern Colorado

Megan Maddocks
University of Northern Colorado

Samuel Miller
University of Northern Colorado

Hattieson Rensberry
University of Northern Colorado

Blue Rice
University of Northern Colorado

Matthew Rodriguez
University of Northern Colorado

Dean Ryleigh
University of Northern Colorado

Galt Tomasino
University of Northern Colorado

Allison Wheeler
University of Northern Colorado

Authors

Dr. Mohammad Majid al-Rifaie is a Senior Lecturer in Artificial Intelligence (AI) at the University of Greenwich, School of Computing and Mathematical Sciences. He is also a visiting lecturer at Goldsmiths, University of London, where he was a lecturer in computer science for two years, and a Senior Research Fellow at the Faculty of Life Sciences & Medicine, King's College London. He holds a PhD in Computational Swarm Intelligence (CSI) from Goldsmiths, University of London, and since the start of his PhD, he has published extensively in the field, covering both the theoretical grounds as well as the applications of computational swarm intelligence, Evolutionary Computation (EC), Machine Learning (ML), and Deep Neural Networks (DNNs). He has taught in higher education for more than a decade, mostly on subjects relevant to programming languages as well as concepts around the digital form, real-world applications, and philosophical issues around artificial intelligence and the arts. His work in the area has been featured multiple times in the media, including the British Broadcasting Corporation (BBC). Over the past ten years, he has developed a unique interdisciplinary research profile with more than 70 publications, including book chapters and journal and conference papers on CSI, EC, ML, and DNNs, as well as their applications in medical imaging, data science, philosophy, and the arts. In addition to speaking a few human languages, he speaks around a dozen machine languages.

Dr. Anna Ursyn, is a professor and Computer Graphics>Digital Media Area Head at the University of Northern Colorado. She combines programming with software and printmaking media, to unify computer generated and painted images, and mixed-media sculptures. Dr. Ursyn has had over 40 single juried and invitational art shows, and has participated in over 200 fine art exhibitions, including musea, such as over a dozen times at the ACM SIGGRAPH Art Galleries, as well as: Travelling shows; the Louvre, Paris; the NTT Museum in Tokyo (5000 texts and 2000 images representing the twentieth century), and Victory Media Network (the largest moving-image outdoor display, Dallas, Texas, https://victorypark.com/victory-media-network/). Her PhD was on art-science connections visualising processes and products in geology. Her research and pedagogical interests include integrated instruction in art, science, and computer art graphics. She has had articles and artwork published in books and journals. Since 1987, she has served as a Liaison,

Organising and Program Committee member of International IEEE Conferences on Information Visualisation (iV) London, UK, and Computer Graphics, Imaging and Visualisation Conferences (CGIV). She also serves as Chair of the Symposium and Digital Art Gallery D-ART iV. Dr. Ursyn has published seven books and many book chapters. Her artwork was selected to be sent to the Moon by NASA as a part of the MoonArc Project by Carnegie Melon University and travelling shows including Centre Pompidou, Paris. Her work in the ABAD exhibition is in the permanent collections of the Museum of Modern Art in New York, and at the Los Angeles County Museum of Art. Website: `Ursyn.com`

Dr. Theodor Wyeld: As a self-taught code hack, I am just the sort of person this book appeals to. In my earlier years, I loved drawing, painting, and building things. Then I got a degree in architecture and ran my own business for a few years. Now I teach interaction design (mobile phone apps, web apps, Java GUIs, etc.) and 3D media studios (animation, VFX, motion capture, and so on). But I was always fascinated by the power of the computer – what could be done with it, how it worked, the way it could be used to manipulate various sources of information and re-present it in graphical ways. For me, it's all about making sense of the world around us, interactively and graphically. I find the computer a great medium for exploring this notion. My latest interest is in 3D printing and artificial intelligence. I really believe, in time, we should be able to instruct computers to do the boring stuff so we can get on with being creative. This is a bit like my experience going from hand drawing perspective images in architecture to using computers to do the 3D modelling part (between the 1980s and 1990s). It meant I could spend more time on the outcomes and less on getting good at setting up the grids for a perspective drawing! For you, it might be letting Photoshop apply a filter to your digital drawing so you can get on with the overall composition, rather than having to learn a new tool. But, to be really innovative, you need to know how the tools work and even how to build your own tools. Then you can really set the world alight! For me, that's what this book aims to do – give you the insights and inspiration to build your own tools, widgets, and apps, creatively.

About the Cover

THE images on the cover are created by the authors. Here is a brief description of each of the images:

Nuclear Reconstruction of *Life*, Mohammad Majid al-Rifaie
This work focuses on the multi-layered communication between two nature-inspired systems: Game of Life (AKA Life) cellular automaton (CA), and a nature-inspired tomographic technique (PART) [2] used in nuclear imaging. As Life generates patterns over time, PART performs a real-time reconstruction of its perception of the generated pattern. Adhering to the constraints of both systems, this work illustrates the process of this dynamically changing, complex communication.

Speeding, Anna Ursyn
Computer program produced three-dimensional, wireframed designs guiding construction as images with codes taking shape as the iconic image of a horse – multiplied, superimposed, transported in order to offer illusions of time and movement.

ascii 332, Theodor Wyeld
Using dynamic data to generate interesting patterns opens up opportunities for expressing creativity in ways not possible using traditional methods. Music is an obvious data source to begin with. Each dimension – pitch, tone, timbre – can be represented with a different signature – shape, colour, opacity. This polar display demonstrates a simple example discussed in this book.

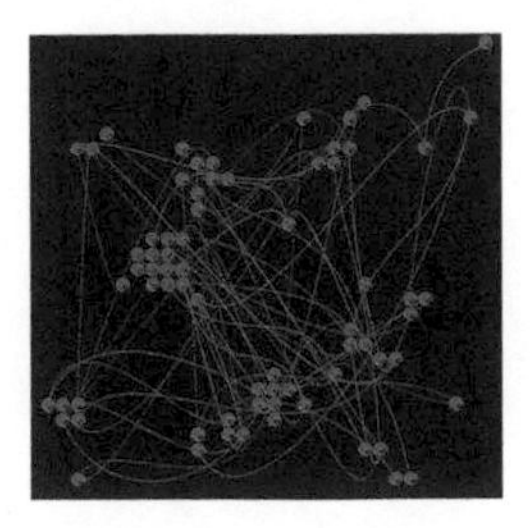
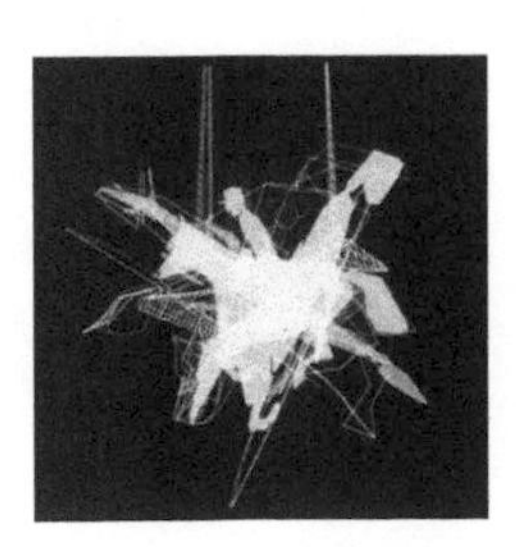
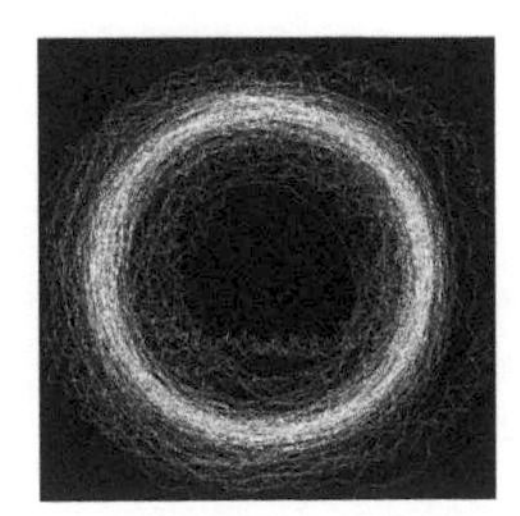

FIGURE 1 Left: Nuclear Reconstruction of *Life*. © Mohammad Majid al-Rifaie. Used with permission; middle: Speeding. © Anna Ursyn. Used with permission; right: ascii 332. © Theodor Wyeld. Used with permission.

CHAPTER 1

Introduction

CONTENTS

1.1 WHO IS THIS BOOK FOR?

In a nutshell, this book covers coding interactions between image and text; music and text; and music and image:

- in the context of programming
- in the context of visualisation
- in the context of media and presenting outcomes on the Internet.

Our aim is to bring programming skills closer to the reader by presenting coding in a visual way. The same task may look and be solved in various ways by coding with different programming languages. Accordingly, this book presents coding from the "bird's-eye view" or perhaps a "plane's view." We can say we learn and work better when enjoying the task. And so this book encourages the reader to experiment, create art by coding, explore their own creative potential through playing, and looking for interactions. It shows software users what coding can offer beyond just knowing the software. In writing this book, we have accessed global contacts, resources, and collaborations. Coding and how to code enables cognitive interest and unleashes the artist in us. We explore various styles in learning with a special focus on project-based approaches with the application of sample code and illustrations.

Visual content can be developed in many ways, in different coding languages. We show several ways of approaching coding for developing visual outcomes, be it using libraries, having a turtle draw the paths, modifying an image, or creating a path that follows some coordinates, and more. We show

how programming can be developed for graphics, 3D content, time based applications and interactivity. For example, in a project involving digital drawing, Myro – a little robot on wheels can draw with a magic marker; a program written by the user triggers this device, and thus Myro follows a trajectory drawn by it according to the code. In another project a user has 10 points to distribute between the robot's intelligence, weaponry, and speed. Depending on the visitor's choice, different robots show up on the screen.

The action of writing a computer program is to instruct a computer what we want it to do. It can take new, previously non-existing forms which assume a new meaning, scope, purpose, role, and techniques. While coding, we use icons driven by metaphors such as a mouse, a desktop, a folder and a brush. New developments in technologies have created new requirements and can enhance the practice of programming in almost infinite ways.

First, we get data by collaborating with other specialists, but some verbal shortcuts invented may cause some communication obstacles, shortcuts that result from personal, technological, and material-related reasons. The multitude of new apps, methodological solutions, and web solutions creates favorable conditions for making shortcuts typical of particular disciplines. Understanding programming can help collaborating partners to convey what they want to share. Shortcuts and accelerated ways of thinking about distinct tasks are reflected in the programming style.

Second, individual programmers may attain the same solution using different programs. While teaching programming, it is beneficial for the learning process to compare a short elegant programming solution with another, more elaborate but perhaps more robust one. While shortcuts may be useful, they must be conducive for exercising creativity, which was identified by the former US President, Barack Obama, as the currency of the 21st century.

Moreover, one notices a new direction in attitude and interest in using a greater number of our senses, in the vivid presence of 3D web, augmented reality, virtual reality, holographic projects, performances, multimedia installations, and so on. Hence, our world becomes more perceptual, especially more visual. This can be seen in the way apps are designed. For example, Apple.com created a foolproof design where only the right connections work.

We need to look at the role played by the humanistic liberal arts, literature, and fine arts in the shaping of the visual/verbal communication in the course of history. We notice a considerable cognitive effort is necessary to comprehend the core concepts related to programming and effort in grasping the puzzles and formulas linked to processing, structuring, and managing information with the use of computing. And so, the goal of this book is also to approach computing and programming in terms of multisensory based communication shared through the networked social environment.

We can say that art media of the 21st century, including music, theater, networked digital media art, and design has been inspired by the input from our senses and includes audiences' response as part of the work. Art and design projects incorporate the viewer's senses in music, theatre, digital media, and art. They often aim to visualise the unseen and give the viewer a phenomenal, immediate experience. Multimedia has since become multisensory after its content and technical solutions merged text, audio, still images, animation, and video footage with interactivity and augmented reality. Generative art created with the use of a computer can be seen as even more multisensory, as it may comprise images, combining painting with digital techniques, computer animations, video, digital music and sounds, videogames, websites, net art, programmed graphics and sculptures, nano art, interactive installations and performances, and augmented or virtual reality events, among other possibilities. Our goal lies in including this world in a sphere of learning science and computing, thus enhancing the possibilities to make it easier and more fulfilling.

As this book's authors are professors coming from different continents (UK, USA, Australia), they may use different teaching strategies when teaching technical problems supported by programming. Thus, in this book, we are sharing our experience of working with students along with our pedagogical, curricular, and conceptual approaches. Our own, some invited coders' and our students' (with their permission) solutions to the same visually oriented tasks are available, along with the code and outcomes.

1.2 CODING: LANGUAGE OR MATHEMATICS?

One of the preventive measures stopping certain groups from coding is dividing pupils at a young age between humanistic or engineering types, something that deters many from acquiring the skills to learn how to code. Programming has long being associated with people active in the fields of mathematics, physics, science and engineering, and it is only in the last few decades that creative or "humanistic" types are finding their way into this world.

One important consideration to take when discussing coding is a fear of mathematics and its link to coding for the "humanistic" person. It is essential to focus on the fact that *learning to code* is, to a large extent, essentially *learning a new language* (a programming language), which has its own grammar, syntax and semantics. Some suggest learning a programming language is like becoming bilingual [55]. It contemplates whether computer programming languages are like literary languages, then investigates whether people who speak one language and can programme to a high standard should be considered bilingual.

There is a large body of literature that argues bilingual people perform faster than the monolingual in tasks requiring executive control; tasks that involve being capable of focusing on the important information and ignoring the irrelevant parts. A more detailed review providing further insights on these issues can be found in [20]. Other researchers have focused on learning a programming language as *learning to solve problems* (i.e. problem solving) rather than a linguistic activity, therefore ignoring the relationship between acquiring a natural language and learning a computer language.

One of the many views held on languages, albeit not universally, is that of Sampson [44], where one's language moulds their perception of reality. In other words, one's world is a linguistic construct. Following from this, Murnane [35] argues that if there is an element of truth in Sampson's views (where language determines, or at least heavily influences, the way human beings think), more emphasis should be placed in investigating the relationship of computing languages and machine languages, particularly with machine languages designed and taught in introductory programming courses.

1.3 CODING WITH VISUAL AND VERBAL CUES

Computing and programming with visual and verbal cues supports studying many fields of knowledge. Skills in programming with the use of computer languages is valuable in creating graphics and art, producing output for web, networking, software, virtual reality, and many other interactive technologies. Our skills in programming allow us to interact with the environment, the machines we use, other people, and materials. For example, Python, Maya, Blender, HTML with CSS and more languages for the web, even software such as Adobe inDesign, Adobe XD for interactivity, Processing, and Arduino are useful for these purposes. Computer programs underlay processes accomplished with 3D printing, along with the field of digital fabrication, allowing production of a whole range of items, including 3D printed pancakes, deserts, and more (at the ACM Siggraph Conference), designing 3D printed sugar laces for the cake, building 3D printed houses [40], or houses that can withstand earthquakes [32, 9]. The list of materials available for 3D printing has been constantly expanding, including plastics, metals, woods, cements, and clay.

Communication seen in a historical perspective comprises visual writings in many modes and styles, visual storytelling, and visual rhetoric. Present solutions include the sharing of messages and knowledge with the use of visual and verbal computing, and other practical applications. The multisensory quality of the web, smart phones, and tablets is changing the format of our daily communications and actions. Computing has become the common trait that defines our frame of reference, both in electronic arts and online social networking sites, where groups of interconnected people exchange information, play and cooperate. The impact of computing technologies can be seen to extend from bio- and evolutionary computing applications, smart and intelligent apps, bots, ubiquitous devices, and smart phones that provide networked exchanges to interactive communications through online social networking. Collaborative projects often begin as kickstarters, such as: Robot Vectors, Vector Art and Graphics[1].

Basic skills in programming can facilitate communication between an artist and a computer scientist, when each side can put into words their needs in a way that is understandable for the other. But, at times, artists are afraid to ask questions, as they are lacking confidence about what they are going to ask, and programmers are lacking confidence about graphical ways of thinking. Thus, artists need knowledge about how a computer works, how a compiler acts and the underlying concepts in programming, while a computer scientist might like to be able to discuss issues related to resolution in the wireframe of a project, or the visual quality of an output. Nowadays many projects in workspaces are increasingly collaborative in nature. Thus, most prospective employers see familiarity with coding as a bridge between coders and visually oriented people, so they can communicate better and faster.

For all these reasons, programming should be taught from early childhood. From the sixties, Seymour Papert insisted that children should use computers as instruments for learning and for enhancing creativity. His vision for early childhood education included the use of programming by children. He developed the Logo programming language and a mobile robot called 'Logo Turtle' to support thinking and problem solving through playing. Papert promoted the 'One Laptop per Child' initiative, and collaborated with the Lego Company in manufacturing the programmable Lego Mindstorms robotic kits, which he considered also useful for professional users. Nicholas Negroponte, a co-founder of the MIT Media Lab developed an XO project, offering a laptop to a child across the globe. People were invited to buy a $100 Linux based the OLPC XO laptop (able to perform most of the basic operations) by paying a double price, so the second machine could go to a child in need [36]. The importance of early learning or early laying of the foundation is not lost on artists or scientists; Pablo Picasso supposedly suggested that to become a great artist one has to preserve the inner child in one's mind.

Resources at the end of this book include website addresses and links for conferences, books, journals, libraries, learning tutorials, discussion boards, bulletin boards, give-and-take community based resources for clipart, textures, boards, art, music, technical support, products, services, web development tools, storage, software applications, and more.

[1]https://www.freevector.com/robot-vectors

CHAPTER 2

Introduction to Coding

CONTENTS

2.1 A BRIEF HISTORY OF CODING

THE first computer is believed to be from the first century C.E. It is called the Antikythera mechanism and is a mechanical computer [39] that showcases the engineering prowess of the ancient Greeks as well as their impressive knowledge of astronomy. Archaeologists believe it was part of a sunken ship carrying it to Rome, and possibly the dark ages ended its further development. More recently, the computer languages began with instructions and data entered via punch cards. Coded in a symbolic language, they used mnemonic symbols to represent instructions. They needed to be translated into machine language before being executed by the computer, so the computer could understand high-level languages that are closer to natural languages. The first programmer was Lord Byron's daughter, Augusta Ada Byron, Countess of Lovelace who is famously known as Ada Lovelace (1815-1852); the programming language Ada was named in her honour. Navy Commodore Grace Murray Hopper (1996-1992) developed one of the first translation programs for the Mark I computer in 1944. The machine code was recorded on a magnetic drum. Computing programming languages have since been used for translations and adaptations of artificial intelligence (AI) systems and genetic algorithms to digital media. However, as noted by Margaret Boden [4], a perfect translation is not a simple matter, both in the case of translating one computer language into another for the sake of artificial intelligence research and in human languages translations: "Even *Please give me six cans of baked beans* will cause problems, if one of the languages codes the participants' social status by the particular word chosen for *Please*" [4, p. 4]. There are two types of translators: The compiler converts the program to low-level languages (machine- and assembly languages) to be executed later; and, the interpreter converts and executes each statement [13].

Only half a century ago, coding was a privilege known and utilised by only a select few, in even fewer institutions and universities. This monopoly has long since gone, and now people as young as primary school children can be taught to code. Furthermore, given the increasingly appreciated significance of coding as an essential skill, countries are now racing to get their coding and programming curriculum up-to-date for children to equip them with this skill as early as possible.

We are living in an era when computers and software are interwoven with society. This can be seen from many perspectives: Artistically, politically, economically and culturally. As such, people from different backgrounds find themselves wanting to express their ideas using creative coding. This includes scientists and scholars, artists, engineers, hackers, bankers, designers, as well as researchers in the humanities and social and political sciences.

The world of programming languages has evolved in terms of allowing for the achievement of more complex results with less effort. Moreover, this progress has not been done at the cost of removing low-level access to the machine. Importantly, from the point of view of this book, it is now possible for anyone to pick a high-level programming language for different tasks, including creative, visual, theoretical, application-oriented, and audio production.

While the history and the stories of the emergence of the various programming languages might be interesting, we will let you look into that. To suffice, a list is included in this book on page 13. It provides an implicit overview of the interconnected nature of the languages and how they benefited from each other. Understandably, there are various claims as to what is the first programming language; the question becomes more challenging to answer when taking into account that there have been some programming languages and advanced features that were never implemented, including Zuse's Plankalkül (meaning Plan Calculus) [15]. Plankalkül was proposed in 1945 but the full description was not published until 15 years after, in 1972. Again the history of how all except one of Zuse's complex machines were destroyed during the war is worth reading. Despite the existing debate in this area, Fortran is widely accepted to be the first compiled high-level programming language.

Interest in Artificial Intelligence (AI) necessitated the emergence of functional programming languages where list processing, as a key process, was added as a feature. In the mid-1950s the interest in AI became apparent amongst mathematicians (e.g. to mechanise intelligence processes such as theorem proving), linguists (interested in natural language processing) and psychologists (who were keen to explore the fundamental processes of the brain). A unifying theme that could potentially address these needs was sought in order to symbolise data in linked lists as computers were mostly dealing with numeric data in arrays. Herbert Simon and Allen Newell were the first to propose the concept of list processing [46] and opened the way towards an exciting new approach, which was later adapted by IBM and bundled with Fortran as the Fortran List Processing Language (FLPL). Later, in 1958, John McCarthy, who was taking a summer position at the IBM Information Research Department, recognised the lack of recursion and conditional expression in FLPL – the only existing list processing language. Therefore, upon his return to MIT in the autumn of that year, he and Marvin Minsky laid the foundation of the second list processing language (LISP).

Since then, dozens of programming languages have been proposed and developed for various purposes. A recommended source for those interested in the evolution of programming languages is "History of Programming Languages" by Richard Wexelblat [52] as narrated by the designers of the languages themselves. For a more historical perspective of the languages up to Fortran, a chapter in Encyclopaedia of Computer Science and Technology by Knuth [19] and a paper titled "Early Development of Programming Languages" by Donald Knuth and Luis Trabb Pardo [27] is recommended.

On a related topic, Figure 2.1 shows a work created by a University of Northern Colorado student, Dean Ryleigh, which illustrates the history of computing in the graphical form of a spiral fractal. A brief history of digital art can be found in Bruce Wands' "Art in the Digital Age" [51].

2.2 CHOOSING A PROGRAMMING LANGUAGE

CHOOSING a programming language is like choosing a holiday destination from a long list of possibilities; it is a difficult task which depends on many factors, including the time of the year, weather, budget, activities

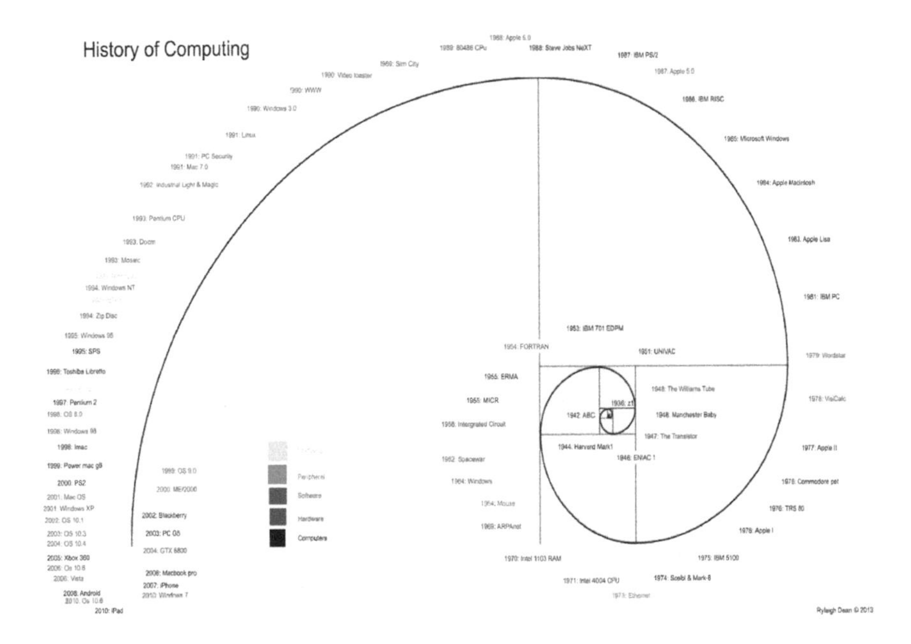

FIGURE 2.1 History of computing. © Dean Ryleigh. Used with permission.

planned and other personal interests. Therefore, different experts offer different recommendations. There are many experts recommending or ranking various programming languages based on their world view and expertise. However, in order to take into account the "global picture" rather than an individual's more limited views, it is worth looking at what languages are used on GitHub, a popular platform for coders allowing millions of coders to share their code in various languages. GitHub provides a good insight on what is going on in the world of coding as it is a place where developers learn about new technologies and new ways of tackling old problems, contribute to what other coders have written, and fulfill their curiosity. At the time of writing this book, GitHub had more than 24 million users from 200 countries working on different projects in 337 programming languages. To get an idea of what the trend is, check the details provided by the company on the top programming languages used in GitHub[1]. At the moment we see that GitHub developers consistently choose JavaScript, Java and Python as their favourite programming languages. The reported trend should be seen only as an indicator of

[1]The company provides an annual report "Octoverse" detailing the most popular programming languages, number of users, projects, etc. To see the full report, visit:

- `https://octoverse.github.com`
- `https://octoverse.github.com/projects#languages`

what GitHub participants are using rather than a guide to choose your programming language, which ultimately depends on your needs and the purpose of programming in the first place.

An alternative ranking suggestion is the Tiobe Programming Index. Tiobe is a software quality company that has been generating an index of the most popular programming languages for decades. Their list is updated monthly based on hundreds of sources[2].

Looking at programming languages from a different perspective, there are often analyses done on the jobs advertised on job search engines such as `indeed.com`, `seek.com.au`, `reed.com.uk` which all host many job postings related to programming. Additionally, you can find the most in demand programming languages on sites such as `techrepublic.com`, `codingdojo.com`, `forbes.com` by simply doing a search on their websites for 'most in demand programming languages'. At the time of writing the book and within one year of checking the top-most in demand programming languages, we noticed a slight reshuffling of some of the programming languages and their ranking. There are lots of reasons for this. Can you think of some?

The story does not end here, and many other institutions provide their own insights into the popularity of various programming languages. For instance, IEEE Spectrum (a magazine edited by the Institute of Electrical and Electronics Engineers) which reports on technology, engineering, and science, has provided another account of a programming languages' popularity where the popularity of Python, C and Java are shown to be almost identical but in varying areas: Python for web applications, desktop developers and hardware programming; followed by C++ for mobile devices, desktop and hardware programming; and Java for web, mobile and desktop applications. The readers are encouraged to check the latest analysis, which also highlights the areas (web, mobile devices, desktops and laptops, as well as hardware) in which the programming languages were used[3]. It is worth noting that in these rankings, 11 metrics from 9 chosen online sources were combined to rank 48 languages[4].

Having gone through some of the statistics, each taking various measures, the main question comes back to: *Which programming language shall I choose?* There is no easy answer to this question, but the good news is that the choice from the vast pool of the programming languages can be narrowed down by specifying the purpose of your coding. Assuming that creative coding is the area of interest, and that you are very new to the concept of programming, there are a few potentially rewarding starting points:

[2]If you are curious about how Tiobe Index is calculated, visit the following: `https://www.tiobe.com/tiobe-index/programming-languages-definition`

[3]`https://spectrum.ieee.org`

[4]The complete interactive list, showing all the programming languages can be found here: `https://spectrum.ieee.org/static/interactive-the-top-programming-languages-2018`

- Processing (based on Java)
- openFrameworks (based on C++)
- Java
- JavaScript
- Python
- PHP
- HTML5
- C++
- C#

They each have their strengths and weaknesses depending on what it is you want to achieve with them. They are all well documented, with a large number of people (both beginners and established digital artists and programmers) using them. Perhaps the easiest to use are Processing and openFrameworks. Processing is a more contained environment (i.e. once Processing is downloaded, you don't need to configure the coding environment). With openFrameworks, some configuration might be required before the result of the first program (Hello World!) can be seen! If you are serious, curious and interested, either one would be fine. There are many resources on the internet for these languages. Hence, in this book, we are trying to stretch you a bit by using other languages in addition to Java and C++. This is based on the collective and individual knowledge derived by observing students, beginners and enthusiasts when exposed to different languages and monitoring their progress.

Having thought about the languages mentioned above, it is now time to briefly look at a list of (almost) all existing programming languages. An alphabetised list of all notable programming languages in existence, both those in current use and historical ones, can be found on Wikipedia[5].

```
    A# .NET, A# (Axiom), A-0 System, A+, A++, ABAP, ABC, ABC ALGOL, ABSET, ABSYS,
ACC, Accent, Ace DASL, ACL2, ACT-III, Action!, ActionScript, Ada, Adenine, Agda,
Agilent VEE, Agora, AIMMS, Aldor, Alef, ALF, ALGOL 58, ALGOL 60, ALGOL 68, ALGOL W,
Alice, Alma-0, AmbientTalk, Amiga E, AMOS, AMPL, AngelScript, Apex (Salesforce.com),
APL, App Inventor for Android's visual block language, AppleScript, APT, Arc, ARexx,
Argus, AspectJ, Assembly language, ATS, Ateji PX, AutoHotkey, Autocoder, AutoIt,
AutoLISP / Visual LISP, Averest, AWK, Axum, Active Server Pages, B, Babbage, Bash,
BASIC, bc, BCPL, BeanShell, Batch (Windows/Dos), Bertrand, BETA, Bistro, BLISS,
Blockly, BlooP, Boo, Boomerang, Bourne shell (including bash and ksh), BPEL, Business
```

[5] https://en.wikipedia.org/wiki/List_of_programming_languages

Basic, C, C--, C++ - ISO/IEC 14882, C# - ISO/IEC 23270, C/AL, Caché ObjectScript,
C Shell, Caml, Cayenne, CDuce, Cecil, Cesil, Céu, Ceylon, CFEngine, CFML, Cg,
Ch, Chapel, Charity, Charm, CHILL, CHIP-8, chomski, ChucK, CICS, Cilk, Citrine
(programming language), CL (IBM), Claire, Clarion, Clean, Clipper, CLIPS, CLIST,
Clojure, CLU, CMS-2, COBOL - ISO/IEC 1989, CobolScript - COBOL Scripting language,
Cobra, CODE, CoffeeScript, ColdFusion, COMAL, Combined Programming Language (CPL),
COMIT, Common Intermediate Language (CIL), Common Lisp (also known as CL), COMPASS,
Component Pascal, Constraint Handling Rules (CHR), COMTRAN, Converge, Cool, Coq,
Coral 66, Corn, CorVision, COWSEL, CPL, Cryptol, Crystal (programming language),
csh, Csound, CSP, CUDA, Curl, Curry, Cybil, Cyclone, Cython, D, DASL (Datapoint's
Advanced Systems Language), DASL (Distributed Application Specification Language),
Dart, DataFlex, Datalog, DATATRIEVE, dBase, dc, DCL, Deesel (formerly G), Delphi,
DinkC, DIBOL, Dog, Draco, DRAKON, Dylan, DYNAMO, E, EarSketch, Ease, Easy PL/I,
Easy Programming Language, EASYTRIEVE PLUS, ECMAScript, Edinburgh IMP, EGL, Eiffel,
ELAN, Elixir, Elm, Emacs Lisp, Emerald, Epigram, EPL, Erlang, es, Escher, ESPOL,
Esterel, Etoys, Euclid, Euler, Euphoria, EusLisp Robot Programming Language, CMS
EXEC (EXEC), EXEC 2, Executable UML, F, F#, F*, Factor, Falcon, Fantom, FAUST, FFP,
Fjölnir, FL, Flavors, Flex, FlooP, FLOW-MATIC, FOCAL, FOCUS, FOIL, FORMAC, @Formula,
Forth, Fortran - ISO/IEC 1539, Fortress, FoxBase, FoxPro, FP, Franz Lisp, Frege,
F-Script, G, Game Maker Language, GameMonkey Script, GAMS, GAP, G-code, GDScript,
Genie, GDL, GJ, GEORGE, GLSL, GNU E, GM, Go, Go!, GOAL, Gödel, Golo, GOM (Good Old
Mad), Google Apps Script, Gosu, GOTRAN, GPSS, GraphTalk, GRASS, Groovy, Hack, HAGGIS,
HAL/S, Halide (programming language), Hamilton C shell, Harbour, Hartmann pipelines,
Haskell, Haxe, Hermes, High Level Assembly, HLSL, Hop, Hopscotch, Hope, Hugo, Hume,
HyperTalk, IBM Basic assembly language, IBM HAScript, IBM Informix-4GL, IBM RPG, ICI,
Icon, Id, IDL, Idris, IMP, Inform, INTERLISP, Io, Ioke, IPL, IPTSCRAE, ISLISP, ISPF,
ISWIM, J, J#, J++, JADE, JAL, Janus (concurrent constraint programming language),
Janus (time-reversible computing programming language), JASS, Java, JavaScript, JCL,
JEAN, Join Java, JOSS, Joule, JOVIAL, Joy, JScript, JScript .NET, JavaFX Script,
Julia, Jython, K, Kaleidoscope, Karel, Karel++, KEE, Kixtart, Klerer-May System, KIF,
Kojo, Kotlin, KRC, KRL, KRL (KUKA Robot Language), KRYPTON, ksh, Kodu, L, LabVIEW,
Ladder, Lagoona, LANSA, Lasso, LaTeX, Lava, LC-3, Leda, Legoscript, LIL, LilyPond,
Limbo, Limnor, LINC, Lingo, LIS, LISA, Lisaac, Lisp - ISO/IEC 13816, Lite-C, Lithe,
Little b, Logo, Logtalk, LotusScript, LPC, LSE, LSL, LiveCode, LiveScript, Lua,
Lucid, Lustre, LYaPAS, Lynx, M2000, M2001, M4, M#, Machine code, MAD (Michigan
Algorithm Decoder), MAD/I, Magik, Magma, make, Maude system, Maple, MAPPER (now part
of BIS), MARK-IV (now VISION:BUILDER), Mary, MASM Microsoft Assembly x86, MATH-MATIC,
Mathematica, MATLAB, Maxima (see also Macsyma), Max (Max Msp - Graphical Programming
Environment), MaxScript internal language 3D Studio Max, Maya (MEL), MDL, Mercury,
Mesa, Metafont, MetaQuotes Language (MQL4/MQL5), Microcode, MicroScript, MIIS,
Milk (programming language), MIMIC, Mirah, Miranda, MIVA Script, ML, Model 204,
Modelica, Modula, Modula-2, Modula-3, Mohol, MOO, Mortran, Mouse, MPD, Mathcad, MSIL
- deprecated name for CIL, MSL, MUMPS, MuPAD, Mystic Programming Language (MPL),
NASM, Napier88, Neko, Nemerle, nesC, NESL, Net.Data, NetLogo, NetRexx, NewLISP,
NEWP, Newspeak, NewtonScript, NGL, Nial, Nice, Nickle (NITIN), Nim, Node.js, NPL, Not
eXactly C (NXC), Not Quite C (NQC), NSIS, Nu, NWScript, NXT-G, o:XML, Oak, Oberon,
OBJ2, Object Lisp, ObjectLOGO, Object REXX, Object Pascal, Objective-C, Objective-J,
Obliq, OCaml, occam, occam-π, Octave, OmniMark, Onyx, Opa, Opal, OpenCL, OpenEdge
ABL, OPL, OpenVera, OPS5, OptimJ, Orc, ORCA/Modula-2, Oriel, Orwell, Oxygene, Oz,
P[2032?][2032?], P#, ParaSail (programming language), PARI/GP, Pascal - ISO 7185,
PCASTL, PCF, PEARL, PeopleCode, Perl, PDL, Perl 6, Pharo, PHP, Pico, Picolisp, Pict,
Pike, PIKT, PILOT, Pipelines, Pizza, PL-11, PL/0, PL/B, PL/C, PL/I - ISO 6160, PL/M,
PL/P, PL/SQL, PL360, PLANC, Plankalkül, Planner, PLEX, PLEXIL, Plus, POP-11, POP-2,
PostScript, PortablE, Powerhouse, PowerBuilder - 4GL GUI application generator from
Sybase, PowerShell, PPL, Processing, Processing.js, Prograph, PROIV, Prolog, PROMAL,
Promela, PROSE modeling language, PROTEL, ProvideX, Pro*C, Pure, Pure Data, Python,
Q (equational programming language), Q (programming language from Kx Systems), Qalb,
QBasic, QtScript, QuakeC, QPL, R, R++, Racket, RAPID, Rapira, Ratfiv, Ratfor, rc,
REBOL, Red, Redcode, REFAL, Reia, REXX, Ring, Rlab, ROOP, RPG, RPL, RSL, RTL/2,
Ruby, RuneScript, Rust, S, S2, S3, S-Lang, S-PLUS, SA-C, SabreTalk, SAIL, SALSA,
SAM76, SAS, SASL, Sather, Sawzall, SBL, Scala, Scheme, Scilab, Scratch, Script.NET,
Sed, Seed7, Self, SenseTalk, SequenceL, SETL, SIMPOL, SIGNAL, SiMPLE, SIMSCRIPT,
Simula, Simulink, Singularity, SISAL, SLIP, SMALL, Smalltalk, Small Basic, SML,
Strongtalk, Snap!, SNOBOL(SPITBOL), Snowball, SOL, Solidity, SPARK, Speedcode, SPIN,
SP/k, SPS, SQR, Squeak, Squirrel, SR, S/SL, Stackless Python, Starlogo, Strand,

```
Stata, Stateflow, Subtext, SuperCollider, SuperTalk, Swift (Apple programming
language), Swift (parallel scripting language), SYMPL, SyncCharts, SystemVerilog,
T, TACL, TACPOL, TADS, TAL, Tcl, Tea, TECO, TELCOMP, TeX, TEX, TIE, Timber, TMG,
compiler-compiler, Tom, TOM, Toi, Topspeed, TPU, Trac, TTM, T-SQL, Transcript,
TTCN, Turing, TUTOR, TXL, TypeScript, Ubercode, UCSD Pascal, Umple, Unicon, Uniface,
UNITY, Unix shell, UnrealScript, Vala, Verilog, VHDL, Visual Basic, Visual Basic
.NET, Visual DataFlex, Visual DialogScript, Visual Fortran, Visual FoxPro, Visual
J++, Visual J#, Visual Objects, Visual Prolog, VSXu, vvvv, WATFIV, WATFOR, WebDNA,
WebQL, Whiley, Windows PowerShell, Winbatch, Wolfram Language, Wyvern, X10, XBL,
XC (exploits XMOS architecture), xHarbour, XL, Xojo, XOTcl, XPL, XPL0, XQuery, XSB,
XSharp, XSLT - see XPath, Xtend, X++, Yorick, YQL, Yoix, Z notation, Zeno, ZOPL
```

2.2.1 Object-oriented programming (OOP)

The languages we are using in this book use what is called Object-oriented programming. Object-oriented programming is really just a way of organising the assets needed to make an application run. It is not exclusive to Java. In this respect Java is similar to the capabilities of C++ OOP. Java includes a set of class libraries that includes system input and output capabilities as well as data types, network support, internet protocols and many standard features needed to build GUIs. All of these features make Java an easier language to learn than some others. It is not an extensive language – although there are many add-on libraries for extending its capability – and does not rely on a lot of hardware or OS (Operating System) dependencies to run. Therefore it tends to be easier to write, compile, execute and debug programs with. In this sense, it is a suitable programming language to learn for creating aesthetic visual applications. In fact, it can even be used to drive robotic functions in digital sculpture. Its object-oriented structure and syntax are very similar to C++; hence it is a good place to start to learn a computer language, because from here you can move on to more complex languages if so desired.

Object-oriented programming is a bit like the way your computer hardware is connected. There are many individual components each with a different function. Together they work to create a larger system. And just like you do not need to know how each individual component works to get the computer working, in an object-oriented program you only need to know that the various objects correctly interact with each other. The various objects need to be able to make the right commands so that they are inter-operable. Sets of objects can be collectively called a *class of objects*. For example, in Java there is a class called Button. There are many types of buttons, and they can be manipulated in many different ways. A button with some text on it is called an *instance* of the button class. It is actually one type of object within the class of objects called buttons. The class defines the features of a button (size, label, colour and so on) and what it does (clickable, changes colour when clicked, etc.). Once the class has been defined in the code, you can create instances of a button. Each button can have a different appearance and behave in different ways.

This way you can reuse the button class many times over by calling it up later in the program. When you write a Java program, a set of classes are constructed. When the program runs, instances of those classes are created and destroyed as needed. The standard Java classes or its library of classes contains most of the functionality needed to create most applications you can think of (maths, arrays, strings, numbers, graphics, networking and so on). But you will still want to create your own classes for specific tasks. However, these will most often be constructed from the already available classes.

2.3 BASIC CONCEPTS IN CODING

THE basic philosophy behind this book pertains to a discussion of how computing and programming with visual and verbal cues might better support studying other fields of knowledge. The authors have attempted to make it easier to comprehend the core concepts in visualising programming by applying pictures, analogies, and metaphors. Visualising the thinking processes in coding, and ultimately generating "tangible" computational outputs are key elements to both improving the coding experience and enhancing the desired outputs. In a nutshell, this is the aim of this book.

It might be true that artists need to know coding for many different reasons. They may seek inspiration in their own art, need a better use of some software packages (for example Blender, Maya, which has access to Python, or Adobe Dreamweaver), they may apply coding that allows for altering a code and supporting visual shortcuts for pre-designed selectors in CSS, or the way the software handles various languages for static and dynamic pages, such as HTML, CSS, LESS, Sass, JavaScript, JSON, PHP, XML, SVG, or even pre-made templates.

Also, since more tasks at any workspace are performed as a group effort (even Adobe Photoshop has sticky notes to allow co-workers to communicate a client's latest demands), artists, designers, and coders could communicate better as a common computer lingo might not create as many barriers.

2.3.1 Syntax and semantics

The idea of a programming language is explored in two broad categories which involve the examination of *syntax* and *semantics*. The syntax of a programming language is related to the structure of the code and its form of expression. On the other hand, the semantics deal with the meaning of the code. Given there are many existing books and materials available on the subject, we skip providing more information on this and instead look at some of the basic and important elements of a program, which are valid for almost all programming languages:

2.3.2 Assignment statements

This process deals with assigning a value (a number, a word, a sentence, an image, etc.) to a *variable*, which is used to store information. Variables are seen as containers that hold information. This information can be referred to or manipulated. For instance:

```
a = 2
```

The assignment statements above would assign the number 2 to a variable called a. This would allow us to see and/or manipulate the content of a later in the program. At the moment the value of a is 2. For instance, if you now execute the following line:

```
a = a + 1
```

the value of a becomes 3. This is because the initial value of a was 2, and the right side of the equal sign is *assigned* to variable on the left side. Therefore, 2+1 is assigned to a and now the value of a is 3. It is important to remember that picking a meaningful name for variables is a good practice.

2.3.3 Sequences

Sequences are one of the main logic structures in programming. They basically mean that the machine goes through each one, one after another, in order. For instance:

```
a = 2
b = 3
c = a + b
print (c)
```

The lines above represent a sequence where two numbers (2 and 3) are assigned to two variables (a and b) and then their contents are added and assigned to a third variable, c. At the end of the sequence, the value inside the container c is 5. The last line of the code above is responsible for printing (i.e. displaying) the value of c on the screen. Without having the first two lines in the code above, the third line which assigns the sum of a and b, would be meaningless. Therefore order is key in a sequence of code lines, which is something a machine expects you to take into account when coding your ideas.

2.3.4 Selections

Selections are, in principle, logical structures which ask a question (e.g. is a$>$ 3?) before executing a statement or a sequence of statements. If the answer to the question is '*yes*', the machine continues to "read" the next line(s), and if the answer is '*no*', they jump to the part of the code which is related to the '*no*' section. For instance:

```
if ( condition )
  // run statements in this part if the condition holds
else
  // run statements in this part if the condition doesn't hold
```

Therefore if the condition holds, the statement(s) below the IF line are executed; otherwise the statement(s) below the ELSE line are run.

2.3.5 Loops

Loops are another important logical structure and as their name suggests, they are responsible for repeating a statement or a sequence of statements several times. For instance, assume that you would like to display the numbers between 1-1000. One way is to repeat calling the *print* function (explained above) one thousand times by writing:

```
print (1)
print (2)
...
print (1000)
```

Another way is to use a loop statement, which has the following basic structure:

```
while ( condition )
  // if the condition is true, the statement(s) here are run
```

In order to display the numbers between 1-1000 using a loop statement, the following can be used:

```
a = 0
while ( a < 1000 )
  a = a + 1
  print (a)
```

In this code, initially, the value of `a` is set to 0. Then, '`while`', one of the commands for creating a loop, is called and it checks if the value of `a` is less than 1000. If that's the case, the machine enters the loop. Inside the loop, `a` is incremented and becomes 1, then 1 is displayed on the screen and the machine goes back to the first line of the loop statements (the line that '`while`' is used).

Now `a` is 1, which is still less than 1000, so the machine enters the loop again and increments `a` by 1, making the value of `a` equal to 2, which is then printed. This process is repeated until `a` is 999. Then, `a` is incremented by 1, becoming 1000 which is then printed and the machine goes back to the `while` loop. It checks if `a` (which is now 1000) is less than 1000. Given the answer is no, the program leaves the loop statement. Note that there are various types of loops, and the `while` loop is only one of the many types. We will see other types later on in the book.

Next we look at a brief introduction to computer architecture and then some language-dependent coding, in Java.

2.3.6 Programming languages & computer architecture

One of the first conceptual computer architectures which has become the basis of today's computers was proposed by John von Neumann[6]. In von Neumann architecture there are three key components:

- Central Processing Unit (CPU)
- Input and Output Units (I/O)
- Memory

In this architecture, an interesting set of communications between these components was envisaged. The input unit is responsible for receiving information from the computer's outside world (via mouse, keyboard, touch screen and so on), and the output unit is tasked to communicate back information to the outside world (via computer screens, speakers and so on). The CPU is in charge of receiving instructions from memory, which contains data and

[6]John von Neumann, or Neumann János Lajos, was working on his electronic machine in a curious place, in the *Institute for Advanced Study* (IAS) in Princeton, New Jersey in the USA, where heavyweight physicists, like Albert Einstein and logicians, mathematicians like Kurt Gödel were frequenting, a place where the heaviest piece of equipment was a piece of chalk, a placed which was known as the Paradise for Scholars, One True Platonic Heaven, where thinking, "pure thinking", could take place. The Institute was a place where "Monster Minds" took refuge to dive deep in their (abstract) thoughts, in peace, quiet and away from the interference of the tangible world.

It was in this place that John von Neumann went on to construct a new species of electronic computer. The Monster Minds, bright thinkers and arrogant scientists of the then One True Platonic Heaven did eventually get rid of the machinery but only after von Neumann died, in 1957 (aged 53). Although they detested his "ugly" electronic machinery, he, also known as Good Time Johnny, was too likeable as a character to get mad at. With his love for jokes, limericks, off-colour stories, fast cars, noise and Mexican food, he could not be hated even by the Institute regulars who made allowances and exceptions purely for Johnny and no one else [42].

instructions. The instructions highlight the type of manipulations required on the data stored in the memory, and the output is sent back to the memory.

Some programming languages (e.g. C, C++) require better understanding of the memory allocations and hardware whereas in some other languages, these details are more abstracted from the programmers (e.g. Java). This book focuses mainly on languages that do not require knowledge of hardware and memory allocations.

One of the primary purposes of a programming language was the automation of tedious tasks or calculations. Hence, automation, which speeds up the process, was introduced to deal with the complex syntax or structure of the early programming languages. While the previous reasons for the "invention" of programming languages is still valid, an emerging reason is to allow more and more people to join the coding community with different motivations. Despite the evolution of programming languages over time and the increasing friendliness of coding in general, there are still rules to be obeyed for machines to understand what we would like them to do.

In term of the computer architecture, once von Neumann architecture became the agreed protocol (i.e. using CPU, I/O and the memory), the main effort was directed at facilitating an easier communication between these components – abstracting as much detail as possible from users and programmers – instead, allowing the focus to be directed towards the key tasks at hand. The history of this abstraction is long, with many scientists and companies involved, each for their own targets and objectives; the richness of this process can also be demonstrated by looking at the long list of programming languages provided earlier.

These attempts, intentionally or otherwise, democratised the process of computer use in general and programming in particular. What once used to be the "game" of the (self-proclaimed) talented and the "geek", became something that less technical individuals, and those in the creative domains, could get involved in more directly. This process allowed the creative and the artists to be engaged directly in the programming process.

2.4 JAVA, SOME BACKGROUND

LET us now discuss in more detail one of the main programming languages used in this book, Java. Developed by Sun Microsystems in 1991, Java is an object-oriented programming language. It is similar to C++ and is a logical language. It was developed to run on anything – televisions, video recording devices, car computers, even refrigerators and washing machine operating systems. But it gained most popularity as a way to add animated content to the internet using applets. Applets are a method for building graphical user interfaces (GUIs) that run in a browser or stand-alone. They are useful for creating animations, games, and other types of interactive multimedia applications that can be accessed via a browser. Another key attribute of Java is that it is platform independent. It will run on any operating system: Windows, Mac, Unix, Linux. This is achieved with the source code and after it has been compiled. The compiled or binary files contain bytecode which is like machine code in that it can instruct the processor what to do directly rather than through the operating system. Java uses an interpreter called a Java Virtual Machine (JVM) to interface between the bytecode and the machine. This allows it to run on any type of machine.

2.4.1 How Java works

Java is a pure object-oriented programming language. Therefore, everything is contained in a class. For example, in its simplest form, when executed, the following class prints Hello World! To the screen. You can simply copy the below into a text file (using Notepad or other text editor), save and change the file extension to .Java instead of .txt (make sure you name your file the same as the class it refers to, HelloWorld.java). Then, in a command window on a machine with the Java Development Kit (JDK, 2019) installed, compile your Java file using the javac command. This will convert your Java text file into a binary class file with bytecode that tells the machine what to do. Then, if you execute `java HelloWorld` in your command window, you should see Hello World! printed to the screen.

Listing 2.1 Hello World!

```
1  class HelloWorld {
2    public static void main(String args[]) { // public means it is visible
         to all other classes
3      System.out.println("Hello World!");
4    }
5  }
6  /* Note: if you copy and paste from this document into a *.txt type
       document and try to compile you may find that the "quotation"
       characters are of the wrong ASCII type. Make sure to use system
       "quotation" characters. */
```

This simple application includes the core features of Java. The program you want to execute is enclosed in a class definition called HelloWorld (the program is enclosed by the curly braces). The program is contained in a routine called `main`. The `public` part just means it is accessible from other classes – even though we only have one class in this file, most files have many classes and often they need to access each other for information. The `static` part creates enough memory to hold the program while it runs. The `void` part means it does not return a numerical value. If it did we would replace `void` with `int` (short for integer). The parentheses next to `main` contain an argument and type. In this case, an array argument of type `string`. It tells the program to generate a `string` (text) and process whatever the operation calls for inside the curly braces. In this case, the operation is to get the system.out (output) to println (print on a new line whatever is in the parentheses) Hello World.

2.4.2 Attributes and behaviour

Every class written in Java has two components: Attributes and behaviour. Attributes refer to the features of an object in a class (appearance, state, type and so on). They include the initial state of the features – whether they are on or off, how many and so on. Each object or instance in the class has its own variable assigned to it. This is because they may be different types of variables (text, number, array and so on). It is these variables that are changed by the behaviours programmed into the class. Behaviour refers to what the object does (start/stop, change rate of change, change type from integer to string etc.). We use different methods to change the function of objects in the class. For example, if we have a class called 'Tree' which we use to draw a tree with, its various attributes might include: Where it is drawn on the canvas, how many branches it has, their length, width and angle relative to each other, number of leaves, and colours of leaves and trunk. The behaviours associated with our Tree class might include the ability to change where it is drawn on the canvas (move its x, y position), number of branches, their length, width and angle, number of leaves, and their colour. We can do this using different methods. A button might turn leaves on or off. Alternatively, we could use a slider to increase or decrease the number of leaves. The same methods could be used to change the length of the branches, colour of the trunk and so on. We need to define these methods in our class file. Methods have four basic parts: A name, the primitive type or object returned, a list of parameters and the body of the method's definition.

```
1  returnType methodName (type1 arg1, type2 arg2, type3 arg3..) {
2  ...
3  }
```

The returnType can be a primitive type, a class name or void, if it does not return a value. If the method returns an array the square brackets come straight after the returnType. The method's parameters are a set of variable declarations, separated by commas, inside parentheses. The parameters are local variables within the body of the method, whose values are passed when the method is called. The body of the method contains statements, expressions, calls to other objects, conditionals, and/or loops. As long as the method is 'real' – not void – it must return a value. For example the following class defines a makeRange() method. The method takes two integers (lower and upper bounds) and creates an array that contains the integers between them.

Listing 2.2 Range class

```
1  class RangeClass {
2    int[] makeRange(int lower, int upper) {
```

```
3        int arr[] = new int[(upper - lower) + 1];
4
5        for (int i = 0; i < arr.length; i++) {
6          arr[i] = lower++;
7        }
8        return arr;
9      }
10
11     public static void main(String arg[]) {
12       int theArray[];
13       RangeClass theRange = new RangeClass();
14       theArray = theRange.makeRange(1, 10);
15
16       System.out.print("The array: [");
17       for (int i = 0; i < theArray.length; i++) {
18         System.out.print(theArray[i] + " ");
19       }
20       System.out.println("]");
21     }
22   }
```

The output of this program is: `The array: [ 1 2 3 4 5 6 7 8 9 10 ]`. The `main()` method tests the `makeRange()` method by creating a range between the lower and upper bounds 1 and 10. It then uses a `For` loop to print the result to the console.

Java stores arguments in an array of strings which are passed to the `main()` method for execution. The `arg[]` in the `main` method refers to an array of strings called `arg`. The program iterates over the array arguments, handling them as specified inside the `main()` method. Because any arguments passed into the program are converted and stored in an array of strings if they are non-string arguments (numbers, Booleans or characters) then they need to be converted to the right type before returning the results. For example, in the class `SumAverage` below if we want to perform an arithmetic operation on some string arguments we need to convert to primitive types first. We can use the class method in the `Integer` class `parseInt()` to parse our string to integers. It is worth remembering this as any (such as primitive type) arguments passed to an argument array are stored as a string array, therefore for arithmetic operations they need to be converted back to primitive types.

Listing 2.3 Sum and average

```
1    class SumAverage {
2      public static void main(String args[]) {
3        String[] arr = { "1", "2", "3" };// arr is the name we are giving to
             the new arguments
4        int sum = 0;
5
6        for (int i = 0; i < arr.length; i++) {
7          sum += Integer.parseInt(arr[i]);
8        }
9
10       System.out.println("Sum is: " + sum);
11       System.out.println("Average is " + (float) sum / (float) arr.length);
12     }
13   }
```

Next, a simple introduction to animation is provided where a few frames of animations are created along with a morphing exercise.

2.5 INTRODUCTION TO CODING ANIMATION

Animation is the process of manipulating an image or a shape with the aim of giving the impression that an image or shape is moving and/or transforming into a different shape. This process can be followed either manually by hand, or digitally through coding. It is enough to have two frames to create a simple animation (flip/flop).

This section provides a brief introduction to animating objects in two different programs. In the first, several circles are "attached" with a line to a central circle. These objects are then animated. In the second example, the concept of morphing is used to convert a shape into another (e.g. circle to triangle to square and then to a random shape). The programming language/environment/library used here is Processing.org, which were originally developed by Ben Fry and Casey Reas. Processing started as an open source programming language based on Java to help the electronic arts and visual design communities learn the basics of computer programming in a visual context. Therefore, it is similar to Java but in some ways easier to use, and it is more tailored for creative and visual outcomes[7]. First, we would like to create a simple animation where a circle expands as time goes by.

2.5.1 First animation: Expanding Circle

Once Processing is downloaded and run, you will see an editor where you can write your first code. Let's call it `ExpandingCircle` (see Fig 2.2). Processing codes contain two main blocks: `setup()` and `draw()`. See the code for the first animation below:

Listing 2.4 ExpandingCircle

```
1  void setup() {
2    size (600, 600);
3  }
4
5  void draw() {
6    ellipse(width/2, height/2, frameCount, frameCount);
7  }
```

The `setup()` part is responsible for setting up the environment and all the necessary adjustments. In the case of our first code, we just set the size of our display window to 600 pixels wide and 600 pixels high. Then the `draw()` part, works like a loop (i.e. running and re-running line 6 forever), therefore the

[7]To run the code and add more features, you can download Processing from `https://processing.org/` and you can read the Getting Started part as well: `https://processing.org/tutorials/gettingstarted/`.

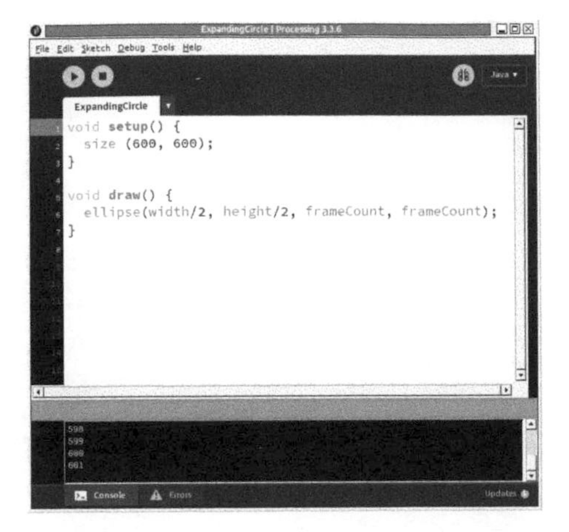

FIGURE 2.2 First animation code. © Mohammad Majid al-Rifaie. Used with permission.

circle grows until it covers the entire display window, and beyond. In line 6, we draw a circle using the `ellipse` method. Here is the *signature* of the ellipse function: `ellipse(a, b, c, d)`, where a is the x-coordinate of the ellipse; b is the y-coordinate of the ellipse, and c and d are the width and height of the ellipse respectively. In our code, the centre of the circle is set in the middle of the display window (`x = width/2`, and `y = height/2`), therefore, (x,y) = (300,300), and the width and height of the ellipse (or the diameter of the circle) are set to the `frameCount` which is responsible for counting the frames, starting from 1 and increasing. Therefore, the animation starts from a circle with diameter 1 and grows without stopping.

In case we want to stop the animation when the circle reaches the edge of the display screen, a conditional statement could be added in the draw() method, after the ellipse:

```
void draw() {
  ellipse(width/2, height/2, frameCount, frameCount);
  if (frameCount > width)
    noLoop();
}
```

This conditional statement tells the program, if the diameter (which is set to `frameCount`) is larger than the width of the display window, stop looping (i.e. using `noLoop()` command).

2.5.1.1 Basic animation of simple objects

In this example, the aim is to create a simple animation with one main circle (`mainC`), to which many other circles (array `c` of type `Circle`) are attached by

FIGURE 2.3 First animation output, snapshot after several frames. © Mohammad Majid al-Rifaie. Used with permission.

a string (line). We would like the circles in c to be at random distances from the main circle and adopt random colours each time we run the program.

Then we would like the circles in array c, to show some movements around the main circle and to expand or shrink in size, giving the impression that while they are attached, they have a "life" of their own.

Once we are done with the above, we would like to show that the attached circles are drawn towards a mysterious power on the left of the screen; or that wind is blowing from right to left. They all gradually move in that direction while at the same time still shrinking and expanding (see Figure 2.4).

2.5.1.2 Class Circle

To do this, we create a `Circle` class, which has the following four attributes (x, y, size, colour). The code for `Class Circle` is given below:

Listing 2.5 Class Circle

```
1   public class Circle {
2     float x;
3     float y;
4     float size; // diameter
5     color Color;
6
7     public Circle() {
8       x = random(width);
9       y = random(height);
10      size = random(100);
```

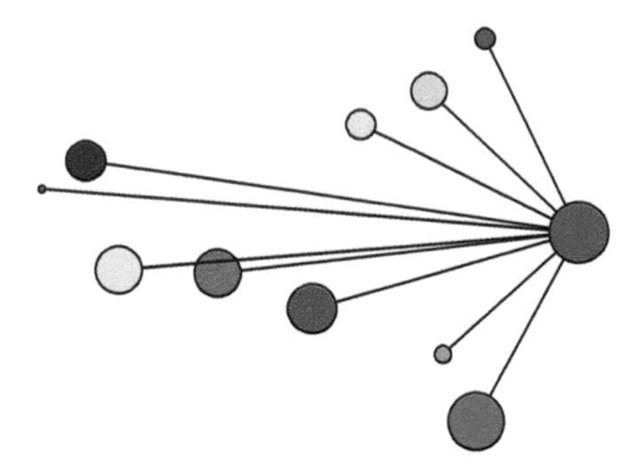

FIGURE 2.4 Moving, expanding and shrinking circles. © Mohammad Majid al-Rifaie. Used with permission.

```
11        colour = color( random(255), random(255), random(255) ); // picking
              random RGB colours
12      }
13
14      // Constructor: taking (x,y) coordinate and size of the circle
15      public Circle(float newX, float newY, float newSize) {
16        x = newX;
17        y = newY;
18        size = newSize;
19      }
20
21      // Constructor: taking (x,y) coordinate, size and colour of the circle
22      public Circle(float newX, float newY, float newSize, color newColor) {
23        x = newX;
24        y = newY;
25        size = newSize;
26        colour = newColor;
27      }
28
29      // drawing the circle
30      public void draw() {
31        fill( colour );
32        ellipse(x, y, size, size);
33      }
34
35      // getters
36      public float getX() {
37        return x;
38      }
39
40      public float getY() {
41        return y;
42      }
43
44      public float getSize() {
45        return size;
46      }
47
48      public color getColor() {
49        return colour;
50      }
51
52      // setters
53      public void setX(float newX) {
```

```
54      x = newX;
55    }
56
57    public void setY(float newY) {
58    y = newY;
59    }
60
61    public void setSize(float newSize) {
62      size = newSize;
63    }
64
65    public void setColour(color newColor) {
66      colour = newColour;
67    }
68  }
```

2.5.1.3 Initialisation and setup()

We can then initialise the main circle (`mainC`) as well as the smaller circles, `c`. The initialisation process determines the coordinate and colour of the main circle, along with the coordinates and size of the smaller circles. Note that the coordinates and the size of the smaller circles depend on the coordinate and size of the main circle.

Listing 2.6 Initialising the circles

```
1   Circle mainC;
2   Circle[] c = new Circle[10];
3   void setup() {
4     size(500,500);
5     background(255); // this sets the background colour to white
6     frameRate(30); // number of frames per second; the larger the number the
          faster the animation
7
8     // initialising mainC with position, size and colour (red)
9     mainC = new Circle(width/2, height/2, 25, color(255,0,0));
10
11    // initialising the circles in c
12    for (int i = 0; i < c.length; i++){
13      c[i] = new Circle();
14      c[i].setX( mainC.getX() + random(-100,100) );
15      c[i].setY( mainC.getY() + random(-100,100) );
16      c[i].setSize( 5 + random( mainC.getSize()-10 ) );
17    }
18  }
```

2.5.1.4 Animation process and draw()

Once the circles are initialised, the next step is to call the `draw()` method and get the animations started. See the code below and the comments provided for each part of the code. Important note: As opposed to standard geometry where the values along the y axis increase when we move up, in computer graphics, the value of y increases when we move down the Cartesian system. This is because the set-out for our screen panel begins in the top left corner and goes down and across.

Listing 2.7 Animating the circles

```
1  void draw() {
2    background(255); // this keeps cleaning the background before showing
         the new frame
3
4    //mainC.setX( mainC.getX() + 0.25 );
5
6    // moving the cicles in c
7    for (int i = 0; i < c.length; i++){
8      c[i].setX( c[i].getX() + random(-3,1) );
9      c[i].setY( c[i].getY() + random(-1,1) );
10     c[i].setSize( c[i].getSize() + random(-1,1) );
11   }
12
13   // draw lines connecting the circles in c to mainC
14   // and draw the circles in c
15   for (int i = 0; i < c.length; i++){
16     line(mainC.getX(), mainC.getY(), c[i].getX(), c[i].getY() );
17     c[i].draw();
18   }
19
20   // draw mainC
21   mainC.draw();
22
23 }
24
25 void keyPressed() {
26   if ( key == 's')
27     save("SampleSnapshot.png");
28 }
```

The full-code (containing the `setup()`, `draw()` methods, along with class `Circle` is provided in the book website.

2.5.1.5 Further coding

After running the code and experimenting with it, you can add more functionalities to the existing program; here are some ideas to get you started:

- Make the main circle move around in the screen
- Then make sure that the main circle does not leave the screen and instead bounces back when it hits the edges
- Relocate one of the small circles by clicking on the canvas
- Put the main circle and the rest of the circles inside a Class and then draw many main circles with smaller circles attached
- Continue, as there is no limit to your creativity!

Also check what happens when you do not clear the background before every frame. You may like the result.

The work in this part, can be seen as a starting point to make more complex animations, where objects, shapes and patterns could be more sophisticated

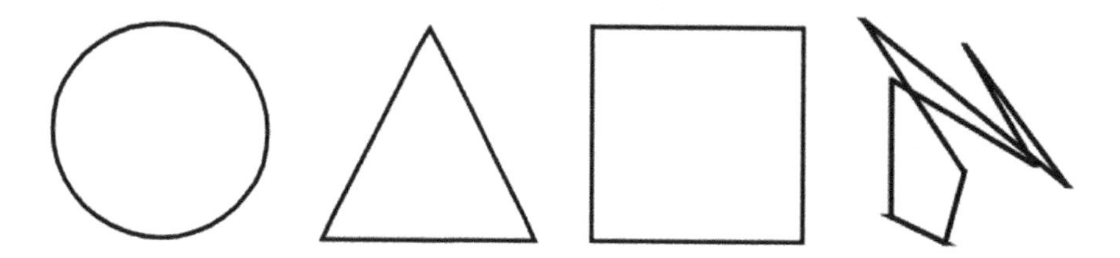

FIGURE 2.5 Morphing a circle into a triangle, into a square, and into a random shape. © Mohammad Majid al-Rifaie. Used with permission.

FIGURE 2.6 Morphing, shapes in between. © Mohammad Majid al-Rifaie. Used with permission.

than a simple circle; movements could be organised with certain trajectories or chasing other visible or invisible goal/s.

2.5.2 Morphing

In simple terms, morphing is the process of transitioning one image or shape into another in a seamless manner. In this example[8], we would like to morph a circle into a triangle first and then into other shapes (see Figure 2.5). In order to achieve this, we need to provide key points of the first object (e.g. circle) and the key points of the second object (e.g. triangle) and so forth. The number of key points should be the same for each object; this would allow each point in the first object to transition into the corresponding point in the second object. Figure 2.6 shows the in-between states.

2.5.2.1 Circle

Initially we create 40 points in a circle, therefore creating a point every 9 degrees (360 / 40 = 9).

Listing 2.8 Circle

```
1  // Create a circle using vectors pointing from center
2  for (int angle = 0; angle < 360; angle += 9) {
```

[8]This section is extended from the code provided in https://processing.org/examples/morph.html.

```
3     PVector v = PVector.fromAngle(radians(angle-90)); // Starting from the
          top centre (therefore angle-90) to match the top tip of the triangle
4     v.mult(50); // scaling the circle (i.e. diameter: 100)
5     circle.add(v); // adding vertix v to the ArrayList circle
6   }
```

2.5.2.2 *Triangle*

We then create 40 points for a triangle. To do this, we could create 10 points to 2 sides each, and another 20 points for the horizontal side of the triangle. In this morphing exercise, we would like the shapes to be between -50 and 50 along the x-axis and between -50 and 50 along the y-axis.

Listing 2.9 Triangle

```
1   // Triangle consists of 3 sides, each of which is drawn below
2   float var = 0; // creating a variable which initially contains 0
3   // The FOR loop below creates the side from (0,-50) to (50,50)
4   for (int i=0; i<10; i++) {
5     triangle.add( new PVector(0+var/2, -50+var) );  // creating vertices for
          the diagonal line on the right-hand side
6     var += 10; // adding 10 to var
7   }
8
9   // The FOR loop below creates the side from (50, 50) to (-50, 50)
10  var = 0;
11  for (int i=0; i<20; i++) {
12    triangle.add( new PVector(50-var, 50) );  // creating vertices for the
          horizontal line
13    var += 5;
14  }
15
16  // The FOR loop below creates the side from (-50, 50) to (0, -50)
17  var = 0;
18  for (int i=0; i<10; i++) {
19    triangle.add( new PVector(-50+var/2, 50-var) );  // creating vertices
          for the diagonal line on the left-hand side
20    var += 10;
21  }
```

2.5.2.3 *Square*

As an alternative we can use a square. A square is formed of four sides, and the code below creates vertices (points) along each of the sides (10 each, therefore $4 \times 10 = 40$). Remember that the total number of vertices for each shape (here, a square) should be the same as the other shapes (i.e. 40 vertices).

Listing 2.10 Square

```
1   // Top side
2   for (int x = -50; x < 50; x += 10) {
3     square.add(new PVector(x, -50));
4   }
5   // Right side
6   for (int y = -50; y < 50; y += 10) {
```

```
7     square.add(new PVector(50, y));
8   }
9   // Bottom
10  for (int x = 50; x > -50; x -= 10) {
11    square.add(new PVector(x, 50));
12  }
13  // Left side
14  for (int y = 50; y > -50; y -= 10) {
15    square.add(new PVector(-50, y));
16  }
```

2.5.2.4 *Random shape*

And to create a random shape with 40 vertices, the following function can be used:

Listing 2.11 Random shape

```
1  void createRandomShape() {
2    for (int i=0; i < 40; i++) {
3      float x = random(-50, 50);
4      float y = random(-50, 50);
5      random.add(new PVector(x, y) ) ;
6    }
7  }
```

If you would like to limit the number of points in the random shape, it is possible to create only a few points (e.g. 8 points) and duplicate them 5 times to get 40 vertices. See the code below:

```
1  void createRandomShape() {
2    // Define random shape with 8 vertices (duplicate each 5 times to have
         40 vertices)
3    for (int i=0; i < 8; i++) {
4      float x = random(-50, 50);
5      float y = random(-50, 50);
6
7      for (int j=0; j < 5; j++)
8        random.add(new PVector(x, y) );
9    }
10 }
```

2.5.2.5 *Animating the morphing process*

Once we have all the vertices for all the shapes, we can start with the `draw()` method in order to lerp (interpolate) each vertex in one shape to the corresponding one in another shape. Note that we have created an empty `ArrayList` of `PVector` (called `morph`) whilst setting up in the `setup()` which will be used in the `draw()` method. The complete code is provided below:

Listing 2.12 Morphing

```
1  // Morph
2
3  // Below, a few ArrayLists are created to store the vertices of the shapes
```

```
4   ArrayList<PVector> circle = new ArrayList<PVector>();
5   ArrayList<PVector> triangle = new ArrayList<PVector>();
6   ArrayList<PVector> square = new ArrayList<PVector>();
7   ArrayList<PVector> random = new ArrayList<PVector>();
8
9   // This ArrayList will contain the vertices of what will be drawn
10  ArrayList<PVector> morph = new ArrayList<PVector>();
11
12  /* The variable below shows which shape is to be drawn:
13   * 0: circle
14   * 1: triangle
15   * 2: square
16   * 3: random
17   */
18  int currentState = 0;
19
20  void setup() {
21    size(400,400);
22
23    // Create a circle using vectors pointing from center
24    for (int angle = 0; angle < 360; angle += 9) {
25      PVector v = PVector.fromAngle(radians(angle-90)); // Starting from the
          top centre to match the top tip of the triangle
26      v.mult(50); // scaling the circle (i.e. diameter: 100)
27      circle.add(v); // adding vertix v to the ArrayList circle
28
29      // Here the ArrayList morph is populated with empty PVectors which
          will be populated later
30      morph.add(new PVector());
31    }
32
33    // A square consists of 4 sides, each of which will be assigned equal
        number of vertices
34    // Top side
35    for (int x = -50; x < 50; x += 10) {
36      square.add(new PVector(x, -50));
37    }
38    // Right side
39    for (int y = -50; y < 50; y += 10) {
40      square.add(new PVector(50, y));
41    }
42    // Bottom
43    for (int x = 50; x > -50; x -= 10) {
44      square.add(new PVector(x, 50));
45    }
46    // Left side
47    for (int y = 50; y > -50; y -= 10) {
48      square.add(new PVector(-50, y));
49    }
50
51    // Triangle consists of 3 sides, each of which is drawn below
52    float var = 0; // creating a variable which initially contains 0
53    // The FOR loop below creates the side from (0,-50) to (50,50)
54    for (int i=0; i<10; i++) {
55      triangle.add( new PVector(0+var/2, -50+var) );  // creating vertices
          for the diagonal line on the right-hand side
56      var += 10; // adding 10 to var
57    }
58
59    // The FOR loop below creates the side from (50, 50) to (-50, 50)
60    var = 0;
61    for (int i=0; i<20; i++) {
62      triangle.add( new PVector(50-var, 50) );  // creating vertices for the
          horizontal line
63      var += 5;
64    }
65
```

```
66      // The FOR loop below creates the side from (-50, 50) to (0, -50)
67      var = 0;
68      for (int i=0; i<10; i++) {
69        triangle.add( new PVector(-50+var/2, 50-var) );  // creating vertices
              for the diagonal line on the left-hand side
70        var += 10;
71      }
72
73      // calling a function that creates a random shape
74      createRandomShape();
75    }
76
77    void draw() {
78      background(255); // setting the background colour to white
79
80      // This value keeps the distance of the vertices from the target
81      float totalDistance = 0;
82
83      // Vertex by vertex lerping
84      for (int i = 0; i < 40; i++) {
85        PVector v1;
86
87        // Deciding which shape to lerp to
88        if (currentState == 0)
89          v1 = circle.get(i);
90        else if (currentState == 1)
91          v1 = triangle.get(i);
92        else if (currentState == 2)
93          v1 = square.get(i);
94        else
95          v1 = random.get(i);
96
97        // Get the vertex to draw
98        PVector v2 = morph.get(i);
99        // Lerp to the target
100       v2.lerp(v1, 0.1);
101       // Check distance to target
102       totalDistance += PVector.dist(v1, v2);
103     }
104
105     // If the vertices are close enough, switch shape
106     if (totalDistance < 0.1) {
107       currentState++;
108       if (currentState > 3) {
109         currentState = 0;
110         createRandomShape(); // each time the random shape is generated from
                scratch and thus different
111       }
112     }
113
114     // Start drawing from the centre
115     translate(width/2, height/2);
116     scale(2); // Double the scale of the drawing
117     strokeWeight(2);
118
119     // Start drawing each one of the vertices, joining the vectices
120     beginShape();
121     noFill();
122     stroke(0);
123     for (PVector v : morph) {
124       vertex(v.x, v.y);
125     }
126     endShape(CLOSE);
127   }
128
129   void createRandomShape() {
130     random.clear(); // clear the previous shape
```

```
131      // Define random shape with 8 vertices (duplicate each 5 times to have
             40 vertices)
132      for (int i=0; i < 8; i++) {
133        float x = random(-50, 50);
134        float y = random(-50, 50);
135        for (int j=0; j < 5; j++)
136          random.add(new PVector(x, y) ) ;
137      }
138    }
```

Obviously more shapes could be added, and there is no limit to how complex your morphing animations can become. For example, Figure 2.7 illustrates the following morphing process:

Circle → triangle → square → house → bird → random shape

The code for generating this morphing is provided on the website and all it requires is declaring the relevant `ArrayList<PVector>` variables for each new shape and then adding the corresponding points `(x,y)`. Note that the number of vertices in this code is 40; therefore, when morphing from one shape to the next, the number of points should be the same in each shape. Below is a snippet of the code for populating the `house` variable with vertices.

Listing 2.13 Adding points for a house

```
 1  void createHouse() {
 2    house.clear();
 3
 4    // Adding the points (vertices) one by one
 5    house.add(new PVector(-50,-20));
 6    house.add(new PVector(-50,25));
 7    house.add(new PVector(-5,50));
 8    house.add(new PVector(50,10));
 9    house.add(new PVector(50,-30));
10    house.add(new PVector(30,-50));
11    house.add(new PVector(-30,-50));
12    house.add(new PVector(-5,-10));
13    house.add(new PVector(-50,-20));
14    house.add(new PVector(-5,-10));
15    house.add(new PVector(50,-30));
16    house.add(new PVector(-5,-10));
17    house.add(new PVector(-5,50));
18    house.add(new PVector(-5,-10));
19    house.add(new PVector(-50,-20));
20    house.add(new PVector(-30,-50));
21    house.add(new PVector(-50,-20));
22    house.add(new PVector(-50,25));
23    house.add(new PVector(-40,30));
24    house.add(new PVector(-40,10));
25    house.add(new PVector(-22,16));
26    house.add(new PVector(-22,40));
27    house.add(new PVector(-50,25));
28
29    // filling the rest of required points with the last vertex
30    while (house.size()<40)
31    house.add(new PVector(-50,25));
32  }
```

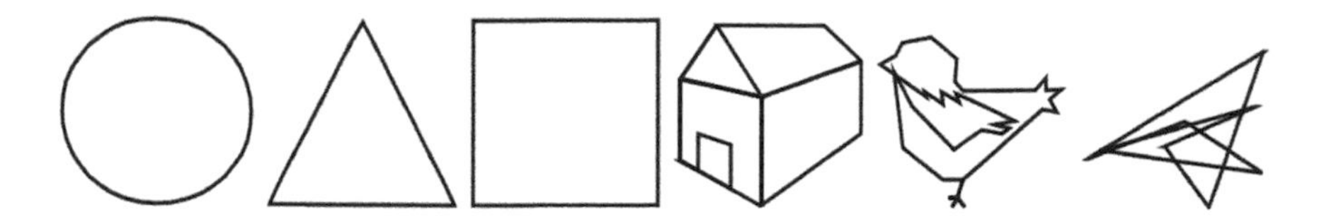

FIGURE 2.7 Morphing a circle into a triangle, into a square, into a house, into a bird, and into a random shape. © Mohammad Majid al-Rifaie. Used with permission.

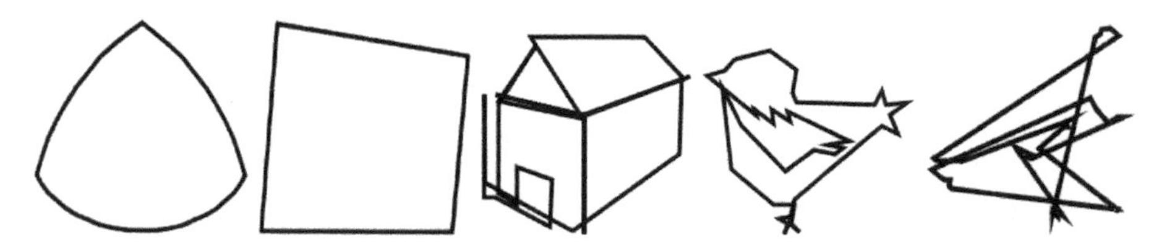

FIGURE 2.8 Morphing, shapes in between. © Mohammad Majid al-Rifaie. Used with permission.

2.5.2.6 Further coding

You can add more features to the code to make the outcome closer to what you have in mind. Here are some ideas to get you started:

- change the number of "corners" in the random shape
- create the vertices of the square in a way that they start from the middle of the top side (i.e. similar to the starting vertex of the circle)
- replace the circle with a pentagon
- "shake" the shapes slightly while then are drawn, creating an impression of shaking objects
- instead of the predefined order in morphing the shapes (i.e. circle, triangle, square, random shape and then again circle), allow the program to randomly switch between the shapes
- create more complex shapes (e.g. cats, dogs, cars, trees)

2.6 AUDIO VISUALISATION

Now we know how to do some Java[9]. What else can we do with it? There is lots of stuff we can do, but, as a creative, making ones' own audio visualiser could be very interesting; as this entails 'seeing' what our favourite music sounds like.

We hear sounds because they make waves in the air due to different pressure ratios, and when these waves reach our eardrums they make the little membrane move causing us to sense sounds. If you have ever taken the grill off a speaker you will notice when music is being played through the speaker you can see the cone moving in and out. The cone is connected to a coil of wire that takes electrical impulses from your amplifier and converts them into the mechanical action of moving your speaker cone in and out. It is these movements of the cone that cause the changes in air pressure to create waves which we detect at our eardrums as sounds or music. We can use the same electrical impulses generated by the amplifier to also change a display on our computer monitor. But we don't use the raw electrical impulses from the amplifier; we can go back to the original source – the CD, mp3, wav or other digital form of the music. By reading in the digital data from the original music source, we can use it to animate some shapes on our computer monitor. That is where the easy stuff ends and the more complicated stuff begins!

The way a speaker cone moves in and out is quite easy to understand – one can see it in action, and it is an analogue signal. By contrast, the digital source of the music is bound up in bits and bytes – which are more difficult to conceptualise as connected with the music sounds they create, when compared to the speaker cone. But, once we can get the sound information data into a form that is more manageable, it starts to make more sense.

There are essentially three components to music: Tone, pitch, and volume. In more technical terms, we can refer to tone as the timbre, quality or character of a specific musical instrument – such as the different tone or timbre produced by a brass or wood instrument; pitch as frequency, cycles-per-second or Hertz (abbreviated to Hz); and, volume as amplitude, level or loudness of a sound. Here, we are only really interested in a sound's pitch and volume. Pitch or frequency refers to how many waves a sound makes in any given second. When we hit water with a stick around 20 times a second we can see the waves created. These waves have a pitch or frequency of 20Hz. This is about the lowest frequency sound that our ears can detect. At the other end of our hearing spectrum the sound of a dolphin squawking is right at the threshold of what is audible and above. On the piano the A4 key has a frequency of 440 Hz, and is often used to 'tune' all the other keys from. While pitch determines

[9]Note that the Processing environment, which was used in the previous section, uses Java as the programming language.

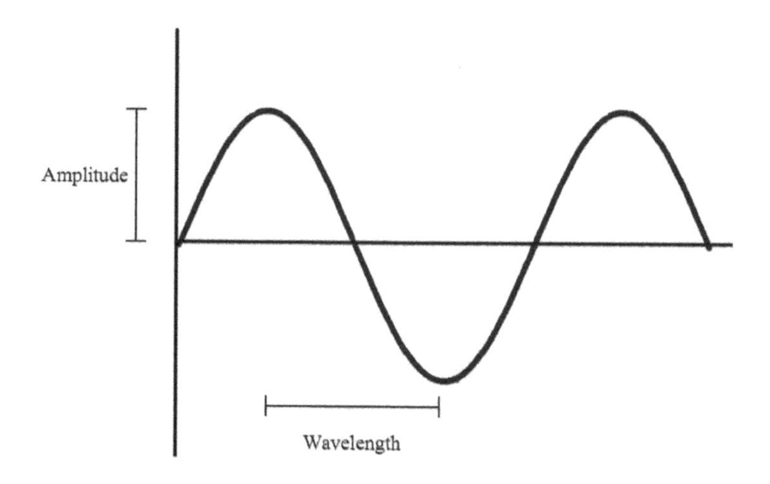

FIGURE 2.9 A sine wave. © Theodor Wyeld. Used with permission.

how quickly the sound waves emanate from the source, volume determines how loud they sound. Volume can be likened to the energy in a sound. A low frequency sound can have high energy – like a jet engine – in the same way a high frequency sound does – like the inaudible sonar a bat uses to navigate at night. The volume and frequency of a sound can be represented on a graph. The volume is represented by amplitude in the y-axis and frequency by the length in x-axis in the form of a wave (see Figure 2.9).

We can track both the frequency and amplitude and use these to drive an interesting visualisation. There are may ways to do this[10]. And, many different languages can be used to support this sort of visualisation (webGL, HTML5 Canvas, Java, JavaScript, C, C++, C#, Python, Processing, and so on). Conceptually, they all follow a similar schema – and some have discrete libraries to do most of the functions – a raw audio signal is analysed and sampled for its amplitude and frequency characteristics. The samples are then used to provide changing values which drive an animation. The animation can be either 2D or 3D. So, let's start by building a 2D audio visualiser.

2.6.1 Building an audio visualiser

In order to build an audio visualiser in Java, we need a method for sampling the audio signal from an audio file (mp3, wav, au etc), make it play (so we can hear it at the same time as seeing the visualisation) and extract the

[10] `http://www.geisswerks.com/milkdrop/`
`https://github.com/robhawkes/webgl-html5-audio-visualiser`
`https://www.airtightinteractive.com/2013/10/making-audio-reactive-visuals/`
`https://github.com/willianjusten/awesome-audio-visualization`

data values we need for our visualisation display (amplitude and frequency). There are lots of different ways to display the audio data after it has been captured and processed. In its simplest form, amplitude can be represented as a continuous linegraph, and, although we could display frequency as a linegraph also, because of the overhead to extract frequency from the audio data, it is simpler to just display discrete bands rather than all frequencies (see Figure 2.10).

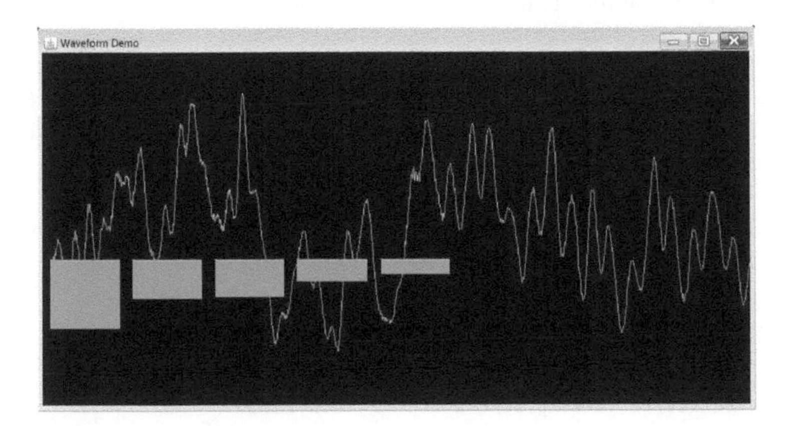

FIGURE 2.10 Basic audio waveform with amplitude visualised as a line-graph and five frequency bands as barcharts. © Theodor Wyeld. Used with permission.

There are three parts to this build: `audioPlayer`, `amplitudeVisualiser` and `volume-frequencyVisualiser`. Each part relies on the previous build.

2.6.2 Basic audio player

The standard Java platform (JavaSE) sound API (in the javax.sound.sampled package), supports the time-sampling[11] of the amplitude of audio files. It supports audio files of type *wav*, *au* and *aiff* (for mp3 files we would need to use the Java Media Framework (JMF), which is not part of JavaSE – so, to keep it simple, we will concentrate on *wav* type files only). The audio sample files can be either 8-bit or 16-bit[12], with a typical sampling rate[13] of 44.1 kHz. Clearly a sampling rate of 44,100 samples per second is way too much data

[11]Time-sampling just means each point of data has two values:

- time from beginning and
- size.

[12]Sample size refers to the number of bits in a sample. Data is stored in 8-bit bytes. Most sound files are not less than 8-bits, with 16-bit typical of CD quality sound.

[13]Sampling rate means the number of times a data stream is sampled per second, usually measured in Hertz or thousands of Hertz (kHz).

to display on a monitor that typically runs at 30-100 frames per second (fps), so it needs to be down-sampled.

The raw audio data is read by Java in 8-bit bytes or samples. To do this the audio data is loaded into a memory buffer. Most audio files have at least two channels (left and right for stereo) and up to 6 for Dolby 5.1. If we are using 16-bit stereo audio files then a frame is a cross-section of samples across the channels in the file. Therefore, for our 16-bit 2-channel file we would have 32 bits per frame [33].

To read the entire audio file we need to create a byte array with the right size; this is done at the part of our code where we read the number of bytes in our sampled data and set our buffer to that size. Then, we can use the number of bytes read to determine the overall sampled data length. Finally, as long as the number of bytes read is greater than or equal to zero we can write or output the dataline as a sound.

```
1    int nBytesRead = 0;
2    byte[] sampledData = new byte[BUFFER_SIZE];
3    while (nBytesRead != -1) {
4      nBytesRead = audioInputStream.read(sampledData, 0, sampledData.length);
5      if (nBytesRead >= 0) {
6        soundLine.write(sampledData, 0, nBytesRead);
7      }
```

Using the `javax` sound package, the `AudioSystem` takes the sound file and prepares it to be analysed. One of the methods for doing this includes reading in lines of data from the audio file. This is like streaming data across the internet – it can be stored, analysed and replayed. To do this we need to start a `DataLine` feed. In the below code you can see we have established a DataLine.Info 'info' from the `SourceDataLine.class`, suitable for processing `audioFormat` type data or files. But, before this, we have identified the `soundFile` with its explicit address and what we are going to do with it: `getAudioInputStream(soundFile)`. We read the entire audio file in using the `AudioInputStream`, then convert the raw data into samples, organised by channel (typically 2 – left and right for stereo). The catches at the end of the class file we have just made are there to let you now if the audio file format you have provided is not supported or a line data feed could not be generated. If you compile and run this class file you should hear the sound from the audio file through your computer's speakers.

Listing 2.14 Basic audio player

```
1  import java.io.*;
2  import javax.sound.sampled.*;
3
4  public class basicAudioPlayer {
5    public static void main(String[] args) {
6      SourceDataLine soundLine = null;
7      int BUFFER_SIZE = 64 * 1024; // 64 KB
```

```
8
9       try {
10        File soundFile = new File("pop.wav");
11        AudioInputStream audioInputStream = AudioSystem
12            .getAudioInputStream(soundFile);
13        AudioFormat audioFormat = audioInputStream.getFormat();
14        DataLine.Info info = new DataLine.Info(SourceDataLine.class,
15            audioFormat);
16        soundLine = (SourceDataLine) AudioSystem.getLine(info);
17        soundLine.open(audioFormat);
18        soundLine.start();
19        int nBytesRead = 0;
20        byte[] sampledData = new byte[BUFFER_SIZE];
21        while (nBytesRead != -1) {
22          nBytesRead = audioInputStream.read(sampledData, 0,
23              sampledData.length);
24          if (nBytesRead >= 0) {
25
26            soundLine.write(sampledData, 0, nBytesRead);
27          }
28        }
29      } catch (UnsupportedAudioFileException ex) {
30        ex.printStackTrace();
31      } catch (IOException ex) {
32        ex.printStackTrace();
33      } catch (LineUnavailableException ex) {
34        ex.printStackTrace();
35      } finally {
36        soundLine.drain();
37        soundLine.close();
38      }
39    }
40  }
```

2.6.3 Basic line graph generator

Now that we have our audio player, we can start to work on how to visualise the data it contains. To do this, we can use a simple line graph. But first we need to know how to do that. So, here is basic linegraph drawing exercise. It draws a line graph based on some random numbers. In the build after this, we will substitute the random numbers for data from an audio file.

Listing 2.15 Simple random point line graph

```
1   import java.awt.Graphics;
2   import java.awt.Graphics2D;
3   import java.awt.*;
4   import javax.swing.JFrame;
5   import javax.swing.*;
6   import java.util.ArrayList;
7   import java.util.Random;
8
9   public class simpleRandomPointLineGraph { // extends JFrame {
10
11    static JPanel view;
12
13    static void drawLines(Graphics g) {
14      Graphics2D g2d = (Graphics2D) g;
15
16      Random rand = new Random();
```

```
17
18      ArrayList<Integer> arrayX = new ArrayList<Integer>();
19      ArrayList<Integer> arrayY = new ArrayList<Integer>();
20
21      int x;
22      int y;
23
24      int temp = 5; // upper limit of random number;
25      // ie generate random numbers between 1 and 5
26
27      for (int i = 0; i < 25; i++)
28      // < 25 is number of times the random number generator iterates (runs
           25
29      // times)
30      {
31        // increment x for each of (i)
32        x = (i) * 20; // offset each x value by 20 pixels
33
34        arrayX.add(x);
35        // add/store the x values to/in our arrayX so we can use them later
36
37        y = Math.abs(rand.nextInt() % temp) + 1;
38        // make y = to the next random integer between 1 and 5 (temp limit)
39        arrayY.add((y * 20) + 50);
40        // because y values are small, mulitply by 20 so they are at least
41        // 20 pixels apart
42        // then add 50 so they are at least 50 pixels down from top of frame
43
44      }
45      g.setColor(Color.RED);
46
47      for (int i = 0; i < arrayX.size() - 1; i++) {
48        // get the arrayX size or length (how many numbers in it)
49        // and let i iterate thsi many times less 1 (so it can start at 0)
50
51        int x1 = arrayX.get(i);
52        // get the first value for i and assign it to x1 on our graph
53
54        int y1 = arrayY.get(i);
55        // get the first value for i and assign it to x1 on our graph
56
57        int x2 = arrayX.get(i + 1);
58        // get the (i + 1), or second, value for i and assign it to x2 on
59        // our graph
60
61        int y2 = arrayY.get(i + 1);
62        // get the (i + 1), or second, value for i and assign it to x2 on
63        // our graph
64        // now, keep doing this for as many times as there are numbers in
65        // our arrayX
66        // and assign their value to the x1, y1 and x2, y2 points on our
67        // graph
68
69        g.drawLine(x1, y1, x2, y2);
70        // draw a line between the x1, y1 and x2, y2 points on our graph
71        // repeat for each iteration of x1, y1 and x2, y2 points on our
72        // graph
73        // until all the numbers in the array have been used up
74
75      }
76      // view.repaint(); //repeats for loop and updates graphic display
77    }
78
79    public static void main(String[] args) {
80      SwingUtilities.invokeLater(new Runnable() {
81        @Override
82        public void run() {
```

```
83
84          JFrame frame = new JFrame("Simple Random-Point Line Graph");
85          JPanel content = new JPanel();
86          frame.setContentPane(content);
87
88          view = new JPanel() {
89            @Override
90            protected void paintComponent(Graphics g) {
91              super.paintComponent(g);
92              drawLines(g);
93            }
94          };
95
96          view.setBackground(Color.WHITE);
97          view.setPreferredSize(new Dimension(500, 200));
98
99          content.add(view);
100
101         frame.pack();
102         frame.setResizable(true);
103         frame.setDefaultCloseOperation(JFrame.EXIT_ON_CLOSE);
104         frame.setLocationRelativeTo(null);
105         frame.setVisible(true);
106       }
107     });
108   }
109 }
```

After compiling and running this program you should see something like that in Figure 2.11.

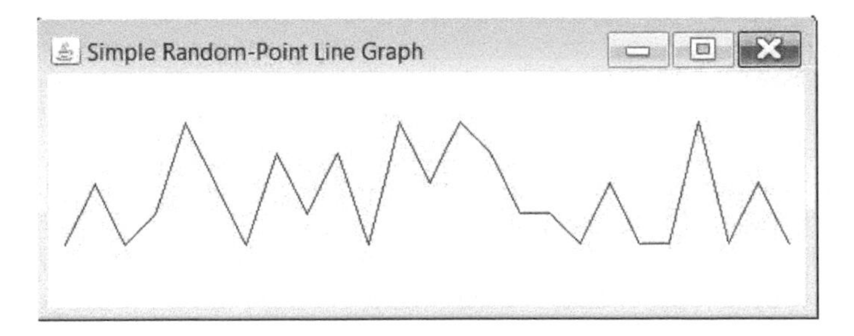

FIGURE 2.11 Random-point line graph.

Now that we know how to create a simple line graph using random numbers for the points, we should be able to substitute the random numbers for values extracted from an audio file. But, just before we build our audio wave visualiser, to demonstrate how we can animate our random-point line graph, try uncommenting `view.repaint();` just before the final curly braces in the `void drawLines(Graphics g)` method. This forces the program to step through the for loop that generates the random numbers again and again, updating the `g.drawline()` which, in turn, updates the view `JPanel`.

2.6.4 Basic audio waveform static visualiser

To build our basic audio waveform static visualiser we need to declare/define our variables, read in an audio file, store a sample, use the samples to draw an image, convert the values into image coordinates to draw a linegraph, initialise a window to display our linegraph on, by calling it from the `main()` method. What we end up with is a static linegraph that represents the changes in amplitude over time of our sound file. In short, similar to the previous build, this visualiser uses an array to store values which are then retrieved as lines between points on a graph. The major difference here is that, instead of generating the points using a random number generator, we are getting our data from an audio file. Whereas in the first audio player build, where we used Java to play a wav file by buffering the data and then processing the data through the PC's speakers, in this build we will intercept the data, scale the values, and store samples used to draw a line graph.

At the top of the code we need to import all the standard Java class files needed for our code to work.

```
1  import javax.swing.*;
2  import java.awt.*;
3  import java.awt.geom.*;
4  import java.awt.event.*;
5  import java.awt.image.*;
6  import java.awt.Point;
7  import java.io.*;
8  import javax.sound.sampled.*;
9  import java.util.ArrayList;
```

At the beginning of our class file we need to declare some variables which we will use later. These include `xStep` which is how many pixels along the horizontal axis before the next point is drawn; `size` holds the dimensions for the window of our panel that we will draw the data on; and, `imageBuffer` is the object that will hold our image information in memory before displaying it on the `JPanel` *view*.

```
1  public class BasicAudioStaticVisualiser
2  {
3    public static final int XSTEP = 4;
4    public static Dimension size = new Dimension(800, 300);
5    public static BufferedImage imageBuffer;
6    public static JPanel view;
7    public static Graphics2D g2d;
8
9    ...}    //this curly brace goes at the very bottom of the code!
10       // it wraps all the other code above into a single class file
11 }
```

Next, we want to load the audio data into a memory buffer using a method that lets us separate out the channels (left and right) and combine the bytes into samples that we can use to generate points on our graph. We do this by reading in the audio `file` as an Audio Input Stream (while we are at it, we will check that it is a valid audio file and throw an exception if it is not):

```
1  static void LoadAudio(String filename)
2  {
3    float[] samples;
4    try {
5      File file = new File(filename);
6      AudioInputStream input = AudioSystem.getAudioInputStream(file);
7      AudioFormat format = input.getFormat();
8
9      if(format.getEncoding() != AudioFormat.Encoding.PCM_SIGNED)
10     {
11       throw
12       new UnsupportedAudioFileException("unsigned");
13     }
```

With the audio file loaded and an Audio Input Stream loaded into memory, we can analyse the file's format. We can find the overall size of the file (in megabits), the byte size of each frame, the sample rate of the original sound recording in kilohertz, and how many seconds the sound file plays for. This information gives us a better understanding of what happens to the file's data as it is being processed by our program. Now we are ready to process the audio data:

Listing 2.16 Processing audio

```
1      long audioFileLength = file.length();
2      int frameSize = format.getFrameSize();
3      float frameRate = format.getFrameRate();
4      float durationInSeconds = (audioFileLength/(frameSize * frameRate));
5
6      boolean big = format.isBigEndian();
7      int channels = format.getChannels();
8      // how many channels in this audio file?
9
10     int bits = format.getSampleSizeInBits();
11     int bytes = bits + 7 >> 3;
12
13     int frameLength = (int)input.getFrameLength();
14     int bufferLength = channels * bytes * 1024;
15
16     samples = new float[frameLength];
17     byte[] audioBuffer = new byte[bufferLength];
18
19     // store/buffer some samples
20     int i = 0;
21     int bRead; //bytes[]  read
22     while((bRead = input.read(audioBuffer)) > -1) {
23
24       for(int b = 0; b < bRead;) { //b here is the variable for bits from
             bytes
25         double sum = 0;
26
27         // (sums to mono if multiple channels)
28         for(int c = 0; c < channels; c++) {
29         // c here is: which channel, Left or Right?
30           if(bytes == 1) {
```

```
31                  sum += audioBuffer[b++] << 8;
32                  // add bytes from each channel
33
34              } else {
35                  int sample = 0;
36
37                  // (quantizes to 16-bit)
38                  if(big) {
39                    sample |= (audioBuffer[b++] & 0xFF) << 8;
40                    sample |= (audioBuffer[b++] & 0xFF);
41                    b += bytes - 2;
42                  } else {
43                    b += bytes - 2;
44                    sample |= (audioBuffer[b++] & 0xFF);
45                    sample |= (audioBuffer[b++] & 0xFF) << 8;
46                  }
47
48                  final int sign = 1 << 15;
49                  final int mask = -1 << 16;
50                  if((sample & sign) == sign) {
51                    sample |= mask;
52                  }
53                  sum += sample;
54              }
55            }
56
57            samples[i++] = (float)(sum / channels);
58          }
59        }
60
61        if(imageBuffer == null) {
62          imageBuffer = new BufferedImage(size.width, size.height,
                BufferedImage.TYPE_INT_ARGB);
63        }
64        //size.width and size.height forces the graphic to fit the panel window
65
66        drawImage(samples);
67        //here we send the samples data to the drawImage method so it can be
              displayed
68
69      } catch(IOException ioe) {
70        System.out.println(ioe.toString());
71      } catch(UnsupportedAudioFileException uafe) {
72        System.out.println(uafe.toString());
73      }
74    }
```

As we know already, the sound file is composed of at least 2 channels (left and right), and the data is stored in bits. We can make sense of how the bits form bytes by organising the flow in a table (see Figure 2.12). Assuming we are using a 16-bit audio file, from the table we can see how each channel contains two lots of 8-bits of information per frame (with 2 channels this is a total of 32-bits per frame). And we know there are 8-bits in a byte. Therefore, we have 2 bytes for each channel frame segment or 4 bytes in total for each frame.

Now that we know how the data is organised, we can work out the total number of frames in a sound file from its overall size. For example, if we have a 2-megabyte, 32-bit per channel, audio file, and we know that 32-bits equals 4 bytes then 2-megabytes$/4 = 500000$ frames. We can also work out how long the sound file should play for. Music is typically recorded at 44.1 kHz. That means

	frame 1	frame 2	frame 3		frame n
left	11010010 • 10111001	01010011 • 00101110	10100111 • 01100110	....	00100011 •10001101
right	10110011 • 00100011	10001101 • 11101001	00101110 • 11101001	....	10111001 • 11010010
	16 bits	16 bits	16 bits		16 bits

FIGURE 2.12 Channels, frames, and bits.

a sample rate of 44100 frames per second. Therefore, if we divide our number of frames by the sample rate we find 500k/44.1kHz = 11.3 seconds duration. The problem with this frame rate is that we cannot possibly display this much information on a screen at 44.1k fps. Therefore we need to down-sample our audio file. A typical fps for displaying animation is 30fps. Hence, for 11.3 secs of sound at 30fps we find we can only display a total of $11.3 \times 30 = 339$ frames. Therefore, we would need to scale our audio data by a factor of $\frac{500k}{339} = 1475$. In other words, we could only display every 1475th frame of our original audio file data – each 1475 frames forms a subset. This is more important later, when we want to display our sound data in real-time. For now, we are only interested in drawing a line graph; therefore we can scale our data to whatever fits the screen. To do this we will average every 'subset' of frames and use this as a new value to draw our line graph – if we didn't average the subsets we might miss some important variations between them – highs and lows.

The processed down-sampled data stream called 'samples' can now be fed to our **drawImage** method. Inside our **drawImage** method we need to normalise the data between the range -1 and 1. This makes it easier to fit our graph parameters (height and width). We can explicitly set our scaling factor by specifying the number of points in our sample that we want to use (**tSamplePoints**). This can further be divided by the set out for our horizontal scale (**xStep**: How far apart each point on the graph will be along the x axis). This gives us our number of subsets we want from samples (**numSubsets**). Finally, we can divide the total number of samples (**samples.length**) by our number of subsets (**numSubsets**) to get our total number of data points used to draw with, and assign this to our variable **subsets**, of type float. We do this by averaging all of the values in a subset and finding its absolute value (non-negative) using **Math.abs**. The values we have now are still pretty big. Hence, if we want them to fit our screen, we need to make them a manageable size. Therefore, the next operation takes the largest and smallest possible values for a 16-bit audio file (± 32768) and scales them between -1 and 1. This is further scaled by the dimensions of the panel window itself.

Listing 2.17 Drawing image

```
1  // use our samples to draw an image
2  static void drawImage(float[] samples)
3  {
4      g2d = imageBuffer.createGraphics();
5
```

```
6         int tSamplePoints = 1000;
7         // total number of points in samples to be used
8
9         int numSubsets = tSamplePoints/XSTEP;
10        // scaled by XSTEP (currently 4)
11
12     int subsetLength = samples.length / numSubsets;
13
14        float[] subsets = new float[numSubsets];
15
16        // find the absolute (Math.abs) average for each subset
17        int s = 0;
18        for(int i = 0; i < subsets.length; i++) {
19            double sum = 0;
20            for(int k = 0; k < subsetLength; k++) {
21                sum += Math.abs(samples[s++]);
22                //add all values within a subset.samples together
23            }
24
25            subsets[i] = (float)(sum / subsetLength);
26            //divide the sum of a subset by its length (how many values in the
                  subset)
27            //and assign it to subsets
28        }
29
30        float normal = 0;
31        for(float sample : subsets) {
32            if(sample > normal)
33                normal = sample;
34                // normal is the biggest number in the sample range
35        }
36
37        normal = 32768.0f / normal;
38        for(int i = 0; i < subsets.length; i++) {
39            subsets[i] *= normal;
40            subsets[i] = (subsets[i] / 32768.0f ) * (size.height);
41        }
42        // now all subset values are between -1 and 1
43    }
```

But before we can actually do the drawing, we need to convert the sample values into image coordinates. To do this we need to retrieve the subset values, store them in an array and feed them to the coordinates of our graph. First we set up a for loop that iterates across all of the subset values (`sample`) assigning them as integers to an array (`arrayY`). We do not really need an array for the `x` values, as these are just increments according to our `xStep`. But, we still need to know what value to assign to the x coordinate for each increment for the total number of the subsets there are. Therefore, we use the `i` to increment `xStep` for the `x` values and store these in `arrayX`. Once we have all our subset values stored as integers in our arrays we are ready to feed them to our draw operation. We set up another for loop that steps through the arrays collecting `x` and `y` values and feeding them to the first and second x and y coordinates of our graph. This operation is repeated until all the values in the arrays are used up. Finally, as the `drawImage` method is displayed on the `view JPanel`, we need to 'repaint' it (`view.repaint();`) to update the actual calculations that occurred after the `drawImage` was first executed.

Listing 2.18 Drawing image, extended

```
1      ArrayList<Integer> arrayX = new ArrayList<Integer>();
2      ArrayList<Integer> arrayY = new ArrayList<Integer>();
3
4      int x;
5      int sample = 0; // to initialise, need to give a seed value
6
7      // convert to image coords and do actual drawing
8      for(int i = 0; i < subsets.length - 1; i++) {
9          sample = (int)subsets[i];
10         x = i * XSTEP;
11         arrayX.add(x);
12         arrayY.add(sample);
13     }
14
15     g2d.setColor(Color.RED);
16     for (int i = 0; i < arrayX.size() - 1; i++) {
17
18         int x1 = arrayX.get(i);
19         int y1 = arrayY.get(i);
20         int x2 = arrayX.get(i + 1);
21         int y2 = arrayY.get(i + 1);
22
23         g2d.drawLine(x1, y1, x2, y2);
24     }
25     view.repaint();
26 }
```

Next we need to build a window to display the results of our `drawImage()` method. We need to make our program runnable – that is, create a thread for the audio data to be fed into our program and for the operations to execute on it. Therefore in our `InitializeWindow()` method we specify a frame, panel and the content to display. Once our 'view' panel is invoked whatever is in the imageBuffer can be unpacked for displaying. The parameters for our panel appear below it: `setBackground` and `setPreferredSize`. The last parameters tells our `LoadAudio()` method what wav file to play.

Listing 2.19 Initialise window

```
1  static void InitializeWindow()
2  {
3      SwingUtilities.invokeLater(new Runnable()
4      {
5          @Override
6          public void run() {
7
8          JFrame frame = new JFrame("Basic Audio Static Visualiser");
9          JPanel content = new JPanel(new BorderLayout());
10         frame.setContentPane(content);
11
12         view = new JPanel()
13         {
14         @Override
15         public void paint(Graphics g)
16         {
17           super.paint(g);
18             if(imageBuffer != null) {
19         g.drawImage(imageBuffer, 1, 1, imageBuffer.getWidth(),
               imageBuffer.getHeight(), null);
20             }
```

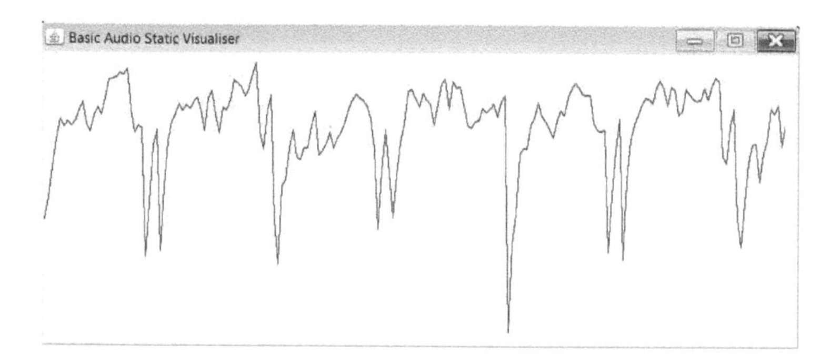

FIGURE 2.13 Basic Audio Static Visualiser. © Theodor Wyeld. Used with permission.

```
21          }
22      };
23
24          view.setBackground(Color.WHITE);
25          view.setPreferredSize(new Dimension(size.width + 2, size.height +
                2));
26      // +2 pixel buffer around image panel
27          content.add(view, BorderLayout.CENTER);
28          frame.pack();
29          frame.setResizable(false);
30          frame.setDefaultCloseOperation(JFrame.EXIT_ON_CLOSE);
31          frame.setLocationRelativeTo(null);
32          frame.setVisible(true);
33          LoadAudio("pop.wav");
34          }
35      });
36  }
```

In the final part of this code build, we need to run our program using a `main` method. This is used to initialise the window which calls all the other elements in the program and loads the audio file which triggers the audio data processing.

```
1     public static void main(String[] args)
2     {
3       // Calling InitializeWindow() Method to Create Window and Populate Data
4       InitializeWindow();
5     }
6
7   }
```

After compiling and running the above code you should see something like Figure 2.13 (depending on the `xStep`, `size`, `tSamplePoints` settings and what audio file you used).

2.6.5 Basic audio animated visualiser

Now that we have our audio player and our line graph visualiser, we can put them together to display the sounds as they are played in a more dynamic, animated way – a static visualiser for audio data is interesting but not very useful. Something that lets us listen to the audio and display its waveform in realtime is more useful.

In this next build we will take an input from our audio file, store it in a buffer and use the down-sampled data to drive an animated linegraph. The main difference between the previous builds and this next one is the way the data is handled. We cannot really call it a real-time visualiser, as the data is stored, processed and replayed. This means there is a delay between when the original audio file is loaded and the information it contains is displayed. Nonetheless, as long as the audio that is played to the speakers is synchronised with the visual display it is sufficient for our purposes.

The main difference is that we are playing the sound at the same time as the line graph is being updated as the line graph is animated rather than just drawn once. In order to do this we need to include in our codebase a method for playing the sounds, `PlaybackLoop` and another method for animating the line graph data, `DisplayPanel`.

There are three states for our playback loop: `loading`, `started` and `done`. `PlayerRef` handles the data flow from the playback loop. It gets the wav file, draws data to the screen and alerts the program when the file is finished.

If we look at the structure of the file we notice that the variables and objects are declared or defined at the top[14]. Making them all `public` means they are available from the various methods and classes within and outside our overall class file. If instead we declared them `private`, they would only be available to this class file. Later we will need to access these variables outside this class file.

```
1  public static final int DEF_BUFFER_SAMPLE_SZ = 2048;
2  public static JPanel view;
3  public static BufferedImage image;
4  public static Path2D path = new Path2D.Float();
5  public static Dimension pref ;
6  public static final Object pathLock = new Object();
7  public static JPanel contentPane = new JPanel();
8  public static File audioFile;
9  public static AudioFormat audioFormat;
```

The first variable determines the size of our data buffer – how much data will be held in memory and displayed on the resulting linegraph – this is not

[14]We can declare a variable such as `int x`; and we can define what it is such as `int x = 1`; and object, or instance of a class might be `JPanel view`; (`JPanel` is the class and we are calling an instance of its '`view`').

the same as the image buffer. Then we need to declare a panel to view the results. The image buffer is used to hold the image in a pixel array which is constantly being updated and drawn to the screen. The actual lines between the points on our graph are paths – a subset of the class awt.geom.Path2D. They are quicker to draw than lines. We use Dimension so that we can set how much is displayed to our screen based on the sample size. The class Object sits at the top of the JavaSE tree of classes. It is used here as an anonymous class so that we can synchronise the audio output (what we hear in the speakers) with the paths drawn (what we see on the screen). The content pane is what we add to our JPanel to draw on. We are using the term File as a container for our audioFile, and AudioFormat so we can access its format type.

Next, we have set up our PayerRef to handle the data and feed it to our display.

```
1  public interface PlayerRef
2  {
3    public File getFile();
4    public void playbackEnded();
5    public void drawDisplay(float[] samples, int svalid);
6  }
```

The next part loads the playerRef and executes its operations – including what to fetch and process (the audio file and send samples to the display method).

Now we are ready to LoadAudio(). As in previous builds, this part retrieves the audioFile from the local machine, checks that it is of the correct type and sends it to the PlaybackLoop method for processing.

The PlaybackLoop() method uses SwingWorker <Void, Void> because it has to handle multiple threads running at the same time. And we want our threads to be systematically organised so they do not get out of step or any data is lost. Notice that the word void[15] inside the parameters for the instantiableSwingWorker uses a capital 'V'. This is because, like the normal use of void, they are placeholders but specific in type to the SwingWorker class. The first Void placeholder represents the type of object returned when the worker has finished working, and the second represents the type of information that the worker will use to update the application with. We could do this using Thread, Runtime or SwingUtilities (invokeLater), but SwingWorker is more efficient. It allows us to do some of the processing in the background. We need to send data to the Event Dispatch Thread (EDT). If we used a single thread for our audio application it would hang as it waits to load the

[15]Void can only return null. Void is used when you do not want to use a class name – although Void is a class in its own right, it is instantiable.

next sample and also when trying to draw to the screen. Using `SwingWorker` we can do the lengthy loading stuff in the background and the shorter tasks in the foreground on a different thread. To do this we need to use the subclass of `SwingWorker doInBackground()` method to perform the background computation. `SwingWorker` takes care of the GUI updates while running the task. The tasks are executed on the Swing EDT thread.

Once our data has been processed we can send it to the draw method – in this case `makePath()`. Samples are fed from the `PlaybackLoop()` method, sorted by channel and averaged. Each point on the path is given a first and second coordinate. These are iterated so that they form a continuous wavy line. They are then sent to the image buffer before being `synchronized(pathLock)` and drawn.

Finally we set up our viewing window – `InitializeWindow()` – and feed it the information to display. This is all executed by the `main()` method at the end of our code.

When this code is compiled and run you should see something like Figure 2.14 (the full code is provided in the book website).

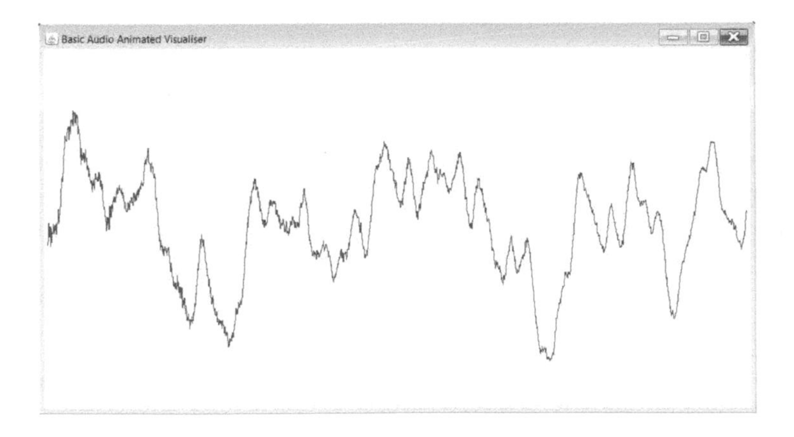

FIGURE 2.14 Output: Basic Audio Animated Visualiser. © Theodor Wyeld. Used with permission.

2.6.6 Basic audio animated visualiser polarised

Now that we know what values to pass to the `geom.Path2D class` to generate a wavy line we can make other shapes also. For example if we want to make our wavy line follow the outline of circle, all we need to do is align the points on the line with those that approximate a circle. In theory, this is like using the y values (causing the up and down movement of the current wavy line) to affect the radius of a circle as it is being swept out. Similarly, we can take the total number of x values and arrange them in a circle by dividing the total number by 360 degrees. A simple formula for performing this function is:

$$x = r.cos\theta \tag{2.1}$$
$$y = r.sin\theta \tag{2.2}$$
$$t = \text{radius of circle} \tag{2.3}$$

To find θ we can use:

$$\theta = 2.\pi/n \tag{2.4}$$
$$n = \text{total number of points wanted on circle} \tag{2.5}$$

But, because our x and y values are set out from $(0,0)$ if we used these formulas as they are, our polar graph would be drawn in the top left corner of the panel. Therefore, we need to add a set-out distance for our x and y values (a and b).

$$x = a + r.cos\theta \tag{2.6}$$
$$y = b + r.sin\theta \tag{2.7}$$

In order to take the points from our audio sample and plot them on this polar graph, we need to first process them through our 'polariser' and then store them in an array. When we are ready to draw our polar graph we can retrieve the new plot coordinates from the array.

Listing 2.20 Plot coordinates

```
1  public static  int mSIZE = 0;
2  // place this at the top of your code with all the other
       declarations/definitions
3
4  // place or make the necessary changes to all of the following
5  // inside the makePath(){ } method
6
7  mSIZE = svalid;
8  ArrayList<Float> arrayAVG = new ArrayList<Float>();
9
10 for(int ch, frame = 0; i < svalid; frame++)
11 {
12     avg = 0f;
13
14     for(ch = 0; ch < channels; ch++)
15     {
16         avg += samples[i++];
17     }
18
19     avg /= channels;
```

```
20
21      path.lineTo((float)frame / fvalid * image.getWidth(), hd2 - avg * hd2);
22
23      arrayAVG.add(avg);
24
25  }
26
27  int a = 250;
28  int b = a;
29  int n =  (mSIZE/2)-1;
30  // n must be (mSIZE/2)-1 as this fits within the buffer sample size
31  // it is divided by 2 because there are 2 channels
32
33  ArrayList<Integer> arrayXx = new ArrayList<Integer>();
34  ArrayList<Integer> arrayYy = new ArrayList<Integer>();
35
36  int k = 0;
37  for (int j = 0; j < n; j++) {
38      double t = 2 * Math.PI * j / n;
39    // if multiplied by more than 2, creates a spiral
40
41      float AVG = arrayAVG.get(k);
42      k++;
43      float r = AVG*200;
44      //to get from normalised ( -1 to 1 ) to big +ve numbers
45    // this multiplication number provides bigger spikes in graph
46
47      int xx = (int) Math.round(a + (r+150) * Math.cos(t));
48      int yy = (int) Math.round(b + (r+150) * Math.sin(t));
49      // the multiplier for r here is to increase the overall radius
50
51      arrayXx.add(xx);
52      arrayYy.add(yy);
53
54  }
55
56  path.reset();
57  // without reset just keeps drawing
58
59  for (int j = 0; j < n - 1; j++) //arrayXx.size()
60  {
61      int x1 = arrayXx.get(j);
62      int y1 = arrayYy.get(j);
63      int x2 = arrayXx.get(j + 1);
64      int y2 = arrayYy.get(j + 1);
65
66      path.moveTo(x1, y1);
67      path.lineTo(x2, y2);
68
69  }
```

After successfully making the necessary changes to the Basic Audio Animated Visualiser, and compiling and running, you should see Figure 2.15. The full code is provided in the book website.

As this is just a graphic image you could apply various filters to achieve different effects (blur, glow, diffuse and so on). If you search online you will find many tutorials on how to apply filters to images in Java. Go ahead and try some different effects to make your own audio visualisation display unique to you.

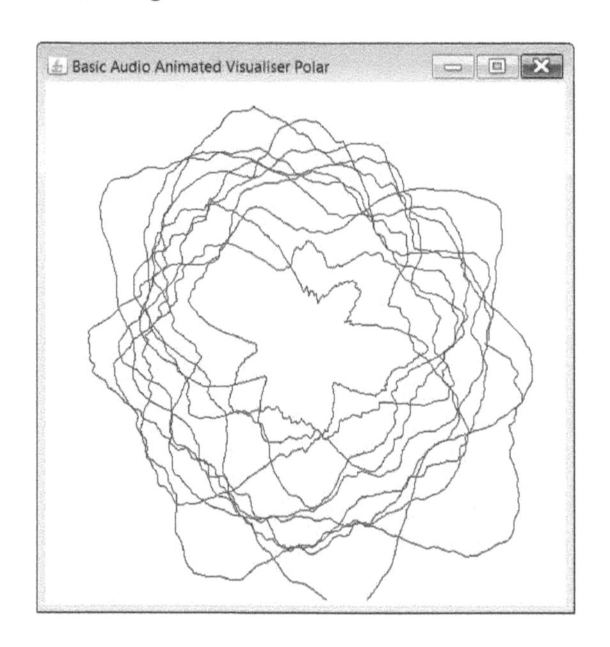

FIGURE 2.15 Output: Basic Audio Animated Visualiser Polar. © Theodor Wyeld. Used with permission.

2.7 MULTISENSORY-BASED PERCEPTION & LEARNING

Just like the coding example in the previous section was about 'seeing' what sound 'looks' like as well as listening to it at the same time, art, graphics, as well as visualisation are all based on our senses. So, we want our 'creative' programming to also work for these domains. Artworks often give us an awareness of our immersion in the surrounding environment. Moreover, advances in the science of generation, detection, and manipulation of light expand our possibilities of sensing and processing signals carried by photons, as well as their interaction with the material world. We capture light and produce photographic or other recordings of movement, frozen in time or moving in films, animations, and movies.

Our multisensory perception of forces acting on senses activates parts of our brain involved in our decision-making. Then comes our response to external forces and internal processes. Hence, a variety of senses and receptors contribute to our communication and multisensory-based learning. The interactive or virtual encounters may refer to varied modes of experience including visuals, the use of light, sound (such as music and voice, including songs), haptic experiences, touch, and gesture. For this reason, discussion of the art of coding should begin from gaining background information about our perception and our reactions to changes in our internal and external environment.

We can say human history has been formed by the ways our various senses were used in the human exchange of ideas and knowledge. We have a multitude of them, much more than the five senses we are taught about in school: Sight, hearing, taste, smell, and touch. Senses give us the capacity to gather data for perception. For each sense we may want to know the kinds of stimuli (physical or chemical; internal or external) to which a particular sense will react, structure of the sense organ and pathways the signals use to access the central neural system, and the resulting reactions of effector organs that act in response to a stimulus, often simultaneously in varied ways.

Our nervous system has specific structures dedicated for each sense that are sensitive to particular signals coming from our environment or from the inside of our bodies. For example, the sensory nerve endings in one's nose have direct connections to communicate with the brain. In addition to the five traditionally recognised senses, our body reacts to signals using several internal and external senses. Our sensory modalities include thermoception reacting to changes in temperature; proprioception – kinesthetic sense perceiving body's position and its motion, and relative movements of the body parts; nociception reacting to pain; equilibrioception supporting balance; mechano reception of vibrations; magnetoception registering directions (that is much stronger in some animals); different kinds of chemoreception (e.g., reacting to concentration of minerals or carbon dioxide in the blood); and a large number of other internal receptors sensitive to processes in internal organs.

Our reactions to stimuli in terms of involuntary reflex actions in response to these stimuli go along a sequence of events mediated as a reflex arc: From receptor organs, then sensory neurons, to nerve centres in the brain or the spinal column, and then, often through interneurons, to motor neurons, and finally effectors such as muscles or glands. Specific stimuli invoke sensual reactions of groups of sensory cells. Signals from these structures activate particular regions in the brain, which interpret them and evoke sometimes complicated reactions from the body. For example, a pupil size is reduced in response to light, also when an object is coming close to the eye, to say nothing about emotional reactions.

Figures 2.16, 2.17, 2.18 and 2.19 show work by students from the University of Northern Colorado picturing organisation of our world: The World in Scale. It is interesting to see how a small part of a visible spectrum can be seen by a human.

The notion of sense is often a compound issue. For example, when we discuss the sense of sight, we may think about several structures reacting to different stimuli, for example, a pupil reacting to changes in illumination, or a retina with its light-sensitive cells: The rods and the cones that allow vision. As for the ear, we can focus on the cochlea that enables the sense

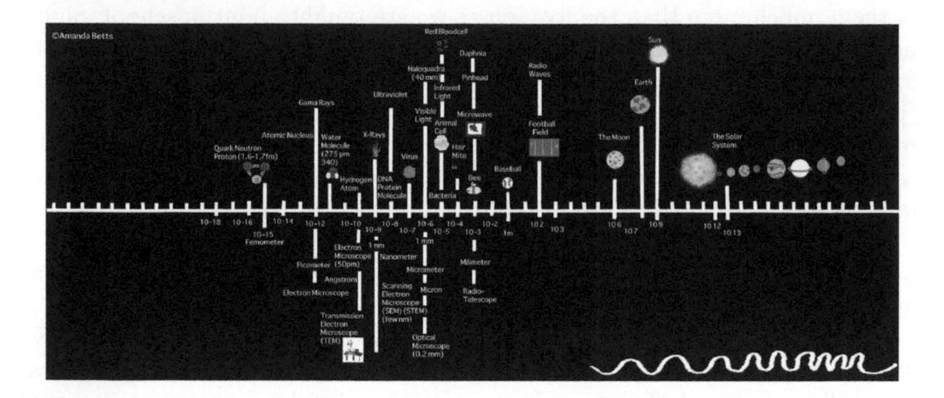

FIGURE 2.16 The world in scale. © Amanda Betts. Used with permission.

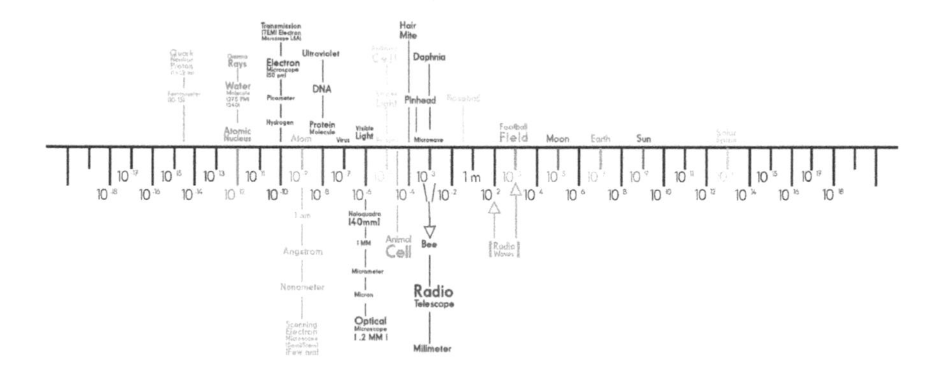

FIGURE 2.17 The world in scale. © Matthew Rodriguez. Used with permission.

FIGURE 2.18 The world in scale. © Allison Wheeler. Used with permission.

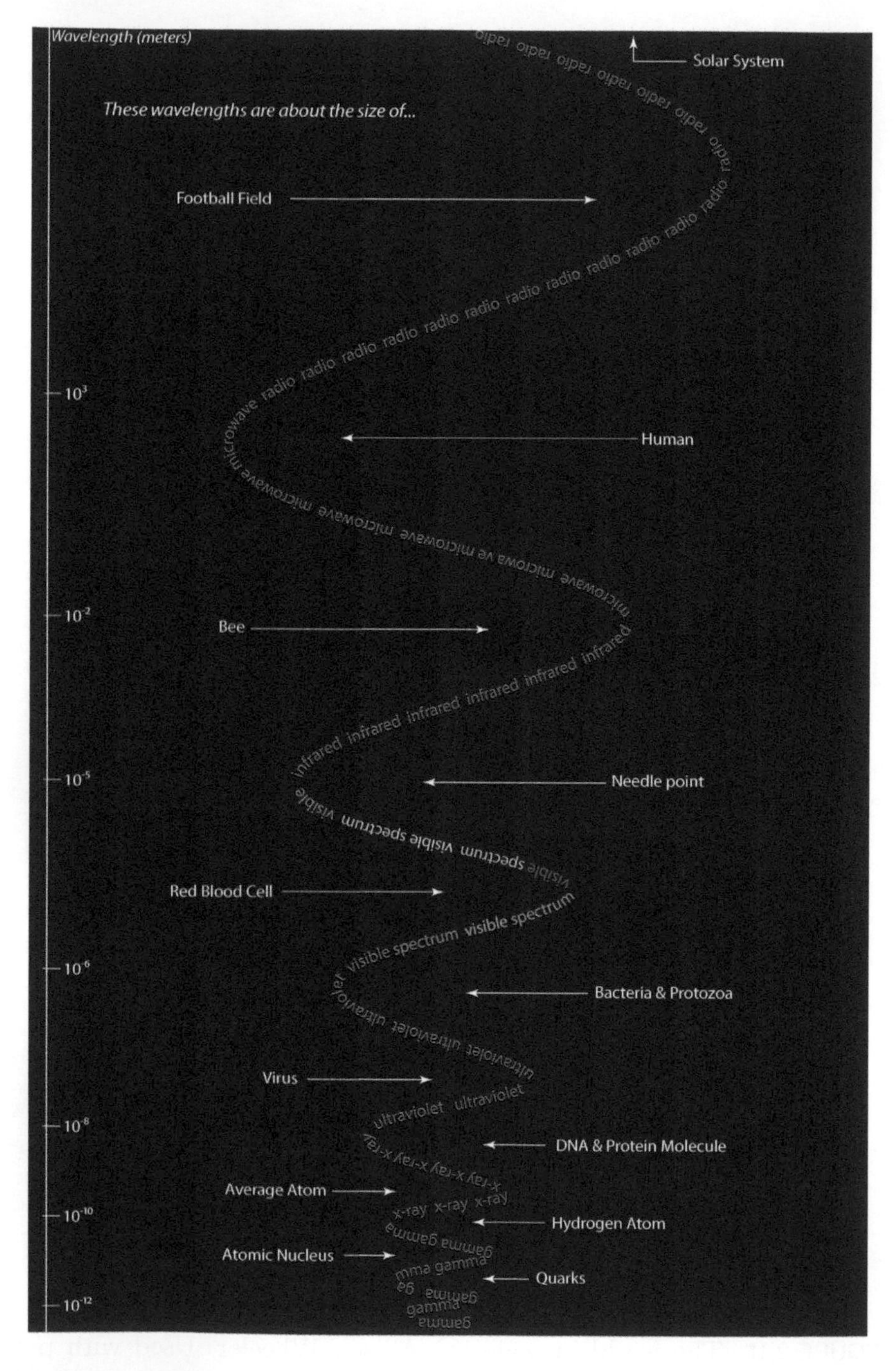

FIGURE 2.19 The world in scale. © Morgan Hurtado. Used with permission.

of hearing, or on other senses such as two ventricles facilitating the static balance and a sense of gravity, as well as on semicircular canals supporting dynamic balance and a sense of acceleration. Sensations of particular tastes or a feeling of hunger may be seen as other compound responses to several stimuli acting on different levels. Signals from the external and internal receptors support maintaining homoeostasis of the whole organism through interdependent physiological processes that keep it in a stable condition.

Moreover, our responses to sensory stimuli are also complicated, as various regions of the brain may respond to any sensory activation. Various sensory modalities are integrated in our central nervous system so they interact with one another. At times they combine muscular, emotional, and psychological reactions, along with some learned, conditioned influences that determine or modify our thinking. For example, verbal associations and preconceptions telling us that blue means sadness may influence our perception of the colour blue. Thus, colour perception may mean activation of several brain structures: Those responsible for emotions of a different kind that may involve visceral reactions, brain structures providing aesthetic experience [56, 57, 58], verbal associations, early memories, or a number of other possible reactions of the mind. Therefore, notions of sense and sensory perception can be seen as working terms that operate within the theme of this book. While coding for a game or an interactive installation one has to consider all these factors. In a similar way, the coding for visual music aimed at sound visualisation involved taking into account visual, acoustical, and psychological factors that are conducive to the poetic or metaphorical translation of music into images and lights, including some widely accepted preconceived principles, for example telling that a higher pitch should be visualised as a lighter colour. Similarly, one can say that a cell phone or Arduino board may react in a similar way to external signals; depending on a programmed threshold, they may react in different ways to the same signal. For example, in swarm intelligence, members of the swarm (think of ants foraging, birds flocking or fish schooling), might react differently from one another despite being exposed to the same stimulus from the environment, yet the collective 'purpose' is to pursue the same objective (e.g. ants: Finding food source; birds: Migrating as a group or flying from my place to the next; fish: Swimming away from predators), which is ultimately achieved.

Learning with multimedia allows a simultaneous involvement of senses in the learning process. Hearing, touch, and other senses beyond word and image, may fortify learning with the use of interactive presentations. One can ponder: 'Is there cooperation or competition between particular senses such as hearing and touch?' Learning materials may also involve studying the cellular level, including factors learned from studying biological systems, genetic impact, swarm inspired computing, and genetics.

Moreover, coordination, and at the same time, competition between word and images may support learning. Below is an essay by the University of Northern Colorado student, Nick Carlson (2014) on the topic [8].

2.7.1 Image versus text

When creating art through writing, painting, drawing, photography, or any other medium, the result is inevitably the same; the work creates an image in the audience's head or gives the viewer a certain perspective towards an issue. The perception of art is mental and each person has a unique interpretation of the work before him or her. However, what allows each medium to be different is the way each attempts to communicate these messages and the limitations that are involved in each medium.

Not only does the content of text matter when working with typography, but also the way those words are presented. What makes language such a beautiful art form, for me at least, is the ability to create entire narratives and concepts through assembling a few words or paragraphs together, allowing for any sort of image to arise and leaving the audience to interpret an image or emotion. However, if the typeface of a work does not match the mood of the content or method of publication, then the audience is left wondering why the artist chose to make that decision, deterring from their reading the image. Therefore, it is important to clarify the purpose of the work and allow for the construction of the typeface to follow accordingly. Truly inspiring works of typography and literature are formulated through the consideration of their font in accordance to their meaning.

What text does with words, images do with texture, placement, and symbolism, which in some cases can greatly strengthen an idea compared to just a description of an idea. Image may clarify a certain idea by giving a physical image to what a person intends, but it could also confuse an audience if not created in a way that communicates a clear idea. Text, in most cases, can deliver intentional, purposeful concepts that images may only lightly touch upon or require background knowledge to understand. However, text and image both require an artist to focus on the colours they use, the shape of the end image, in the case of text how the typeface is arranged, and the proper spacing of the content to allow for easy observation.

2.8 MULTISENSORY PERCEPTION AS A STARTING POINT FOR LEARNING CODING

NEW technologies involve many of our senses, affecting the way we work, communicate, and play. Programming with visual cues may support studying other disciplines. It may be easier to comprehend the core concepts in programming for art, web, and everyday applications by applying pictures, analogies, and metaphors that refer to familiar sensory faculties. Web is a multisensory experience occurring in our social environment. There is a growing need for sharing online visual/verbal messages with emphasis on storytelling and artistic statements. Programming may be seen as a tool giving us the immediate reference to our explorations and actions that are useful both in online social networking and electronic media.

Visual types of programming may involve object oriented programming – a model of language oriented toward objects and data, more than actions or logic. HTML (Hyper Text Markup Language) is a set of markup tags used for describing web pages and building web applications. Markup language is a system for tagging a document to indicate its logical structure (such as paragraphs). Tags replace text providing typesetting instructions. Markups instruct the software about electronic transmission and display of the text [12].

In this book, we will align coding exercises with basic types of art: A portrait, a still life, a landscape, an abstract, followed by interactivity in light of the elements and the principles of design in art.

2.8.1 The elements of design in art

Elements of design refers to what is available for the artist/designer or any person willing to communicate visually. Principles of design describe how the elements could be used [16]. The skilful use of elements and principles of design can enrich the work of art beyond just depiction of reality and invoke aesthetic and intellectual sensations of pleasure, appreciation, or repulsion. Elements of design are also applied for data presentation in all branches of knowledge. In the same way, basic kinds of picturing in art, such as still life, portrait, landscape, and abstract, also obey design elements and principles. "These elements are the materials from which all designs are built" declared Maitland Graves in The Art of Colour and Design [17]. Fine art, functional (utilitarian) pictures, posters, commercials, and all kinds of web productions are evaluated in terms of the elements and principles of design. These elements are known as the fundamentals for all works of art, because without them, art could not be created.

Colour and value, shape and form, space, line, and texture are called the elements of design. All of these elements exist in the world around us in nature and in the environments we create for ourselves.

Depending on the kind of artwork and the message it conveys, particular elements, associated with the specific features to be exaggerated, take over the overall composition. Contrasting colours, distinctive types of line, and a bold use of space are explicitly set forth in politically involved art (such as election related posters), antiwar installations (such as messages displayed in public places), or religious art (such as Russian icons). Furthermore, discourses, emotions, and artistic presentations related to the American flag resulted in a variety of the design solutions. Jasper Johns created several works about the American flag. Big companies' logos were displayed in the place of the flag stars during anti-war manifestations. Some of the manifestations from the nineties included dramatic effects of burning the flag.

2.8.2 Line

A line can be used to demonstrate the shortest path between two points. A line can also be the path to a point which leads the eye through space. Thus, line is a record of movement. It can create the illusion of motion in a work of art. Closed lines define an enclosed space. In a drawing or a painting, line may be used both in a functional and imaginative way and may represent anything: An actual shape, a person, a building, or just an abstract drawing.

Lines can be thick or thin, wavy, curved or angular, continuous or broken, dotted, dashed or a combination of any of these. There are many ways in which we can vary a line in art, by changing the line's width, length, the degree of curvature, direction or position, and/or by altering the texture of the line. The use of line in art involves selection and repetition, opposition, transition, and variety of length, width, curvature, direction, and texture. Movement shown by a line is considered a principle of art.

Norton Juster created "The Dot and the Line: A Romance in Lower Mathematics" [25] using line drawings for his storytelling. The title is considered a reference to a book "Flatland: A Romance of Many Dimensions" by Edwin Abbott Abbott [1]. This book is being constantly reprinted and can be found online, in bookstores, or libraries. "The Dot and the Line" was adapted as a 10-minute animation by Chuck Jones who won the 1965 Academy Award for Animated Short Film. It can be seen online and attracts tens of thousands of visitors. More information and the story of the book can be found on Wikipedia. One can also find a lot of video responses to "The Dot and the Line: A Romance in Lower Mathematics" as online animations of the story[16].

Line in artworks. The ability to perceive line separated from form can become a major aesthetic skill providing an increased awareness of the beauty and function of line. For example, Morris Louis, "Alpha Pi" (1961)[17] shows diagonal parallel lines (acrylic paint) in a Zen-like spirit of meditation. Louis resigned from shape and light in favour of pure colour. El Lisitzky, "Composition" (1920)[18] displays revolving geometric objects painted in colours that appear to be floating in the air, thus creating a sense of depth and an illusion of space. In "Unreal City" by Mario Merz (1968)[19] words are inscribed in neon

[16] For example, `http://www.youtube.com/watch?v=FNXjUsJNiUM&feature=related`

[17] `http://www.metmuseum.org/toah/works-of-art/67.232`

[18] `http://www.wikipaintings.org/en/el-lissitzky/composition`

[19] `http://www.guggenheim-bilbao.es/secciones/programacion_artistica/nombre_exposicion_claves.php?idioma=en&id_exposicion=69`

light within a triangular framework. Is it an elusive and ambiguous metaphor of city life? "Composition" by Piet Mondrian (1929)[20] is built from simple elements – straight lines and primary colours, in search of perfect balance and the order of the universe.

2.8.2.1 Project: Meteorological events

We can use lines creatively, programmatically . Try creating a composition of a wind pattern. First, create a pattern (transform and repeat for your code). Visit the 'Wind' page `hint.fm/wind`, to find an interactive display of real time wind activity in the US. Use transport and repeat function in your code. You may want to get inspiration from maps: Old gravures showing a face in the sky, with the ballooned cheeks blowing into the space from the corner of the picture. You may use principles of design to guide your composition, for example symmetry to show waves of wind or water. Then use dots for rain, snowflakes for snow, etc. Use repetition in code, transform, repeat.

Figures 2.20, 2.21 and 2.22 are some examples of how the students from the University of Northern Colorado used transform and repeat to illustrate how wind dictates the flow of rain, turning into snow.

2.8.3 Texture

Unlike line, texture is a general characteristic for a substance or a material; it exists all around us. A surface can be felt, or is perceived to be felt. Artists add texture to attract or repel viewers' interest regarding an element, depending on whether it is pleasant or repulsing.

Texture can be actual (natural, invented, or manufactured). Tactile texture provides us with a three-dimension feel for a surface, so we can touch it. Painters can use impasto (a thick layer of paint or pigment) to bring about texture on their works.

Texture can also be simulated, made to look rough, smooth, hard or soft, or like a natural texture. Simulated textures are made to represent real textures such as a smooth arm or rough rock formation. But they are not actual textures, and if you touch the picture showing a rough object you feel only the marks of paint, pen, or pencil on a smooth paper. Visual texture is the illusion of the surface's peaks and valleys, like the tree pictured. Any texture shown in a photo is a visual texture.

Try creating a travelling zoo or a circus wagon with animals in it. A rhino or an elephant will have a rough texture, while a skunk will have fur, and a

---

[20] `https://en.wikipedia.org/wiki/Composition_with_Red_Blue_and_Yellow#/media/File`

FIGURE 2.20 Wind, rain and snow. © Allison Wheeler. Used with permission.

FIGURE 2.21 Wind, rain and snow. © Samuel Miller. Used with permission.

FIGURE 2.22 Wind, rain and snow. © Jenny Lee. Used with permission.

porcupine needles. A giraffe will have furry, outlined spots, making a vertical impact on your composition.

2.8.4 Pattern

Like textures, we can see patterns on surfaces. A pattern is an artistic or decorative design made of lines or patches; thus pattern is a repetition of shapes. Patterns make the basis of ornaments, which are specific for different cultures. Owen Jones (1856/2010) made a huge collection of ornaments typical of different countries [24]. He wrote a monographic book titled "The Grammar of Ornament." Textures often have a natural touch but they may have also repeating units or motifs. Regular repetition of a motif may result in a pattern appearing in a texture.

2.8.5 Colour and value

Lines, textures and patterns all have colour. Colour exists almost everywhere. In 1704, Sir Isaac Newton showed how all the colours of the rainbow are contained in white light, such as sunlight. When the light passes through a prism, a band of colours is formed. This band is called a spectrum. Newton also invented a colour pigment wheel. He put the three primary pigment colours: Red, yellow and blue, and the three secondary colours: Orange, green, and violet, in an outer circle. Black is the sum of all of these colours (pigment colours are subtractive, while projected colours are additive). Intermediate colours are the additional hues, which fall between the primary and secondary colours. The mixture of adjoining primary and secondary colours can produce intermediate colours.

A colour wheel: Colour dimensions in pigment

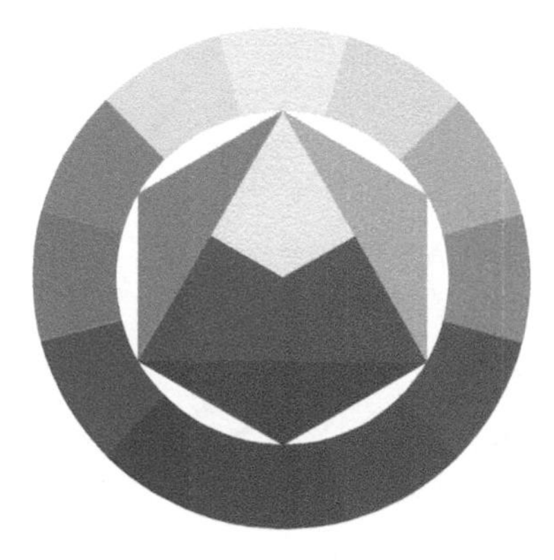

FIGURE 2.23 Shnatsel [CC0], via Wikimedia Commons

Figures 2.24, 2.25, 2.26 and 2.27 are some examples of colours seen as characters created by the University of Northern Colorado students.

The source of colour is light. We see colour because light reflects from the object into our eyes. Light is visible radiant energy made up of various wavelengths. It is one of several electromagnetic waves listed in order of their frequency and length: Long electric waves, radio, television and radar, infrared (felt as heat), visible light, ultraviolet (invisible), X rays, and cosmic and gamma rays.

Colour is a property of the light waves reaching our eyes, not a property of the object seen. The white light of the sun contains all wavelengths of light. When light falls on a surface that reflects all white light, it appears white. When the surface absorbs all the white light, we see the object as black.

Colour in artworks. We can compare the use of colour in paintings. For example, Michelangelo Buonarotti (1475-1564) applied in his work large areas of vibrant, contrasting colours and the play of light over human bodies[21]. Henry Matisse (1869-1954) used in "The Dessert (Harmony in Red)" (1908)[22] primary colours for creating a vibrant, unified pattern of pure colour, painted thickly, without brush marks. The Dutch/American artist Josef Albers (1868-1976) explored the perception of colour both in his paintings and written works, such as *Interaction of Colours* (2010/1963). He had also studied, then

[21] http://www.ibiblio.org/wm/paint/auth/michelangelo/

[22] http://en.wikipedia.org/wiki/File:Matisse-The-Dessert-Harmony-in-Red-Henri-1908-fast.jpg

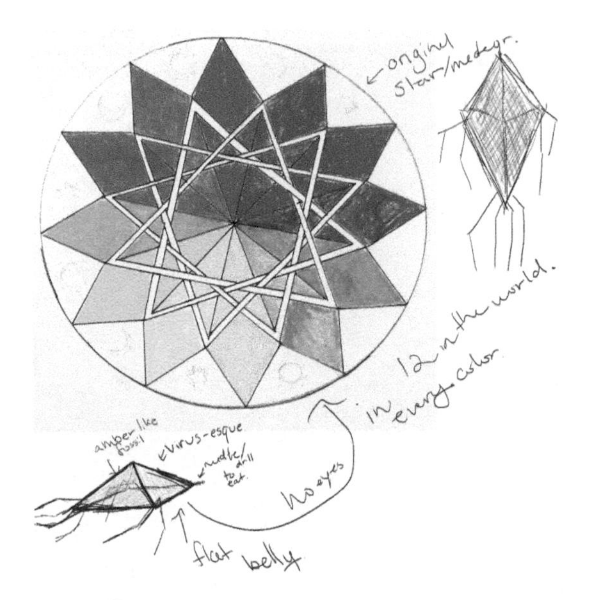

FIGURE 2.24 Colours seen as characters. © Jenny Lee. Used with permission.

FIGURE 2.25 Colours seen as characters. © Hattieson Rensberry. Used with permission.

FIGURE 2.26 Colours seen as characters. © Moises Gomez. Used with permission.

FIGURE 2.27 Colours seen as characters. © Blue Rice. Used with permission.

taught in Germany at the school of architecture and design of the twentieth century, Bauhaus School of Design, which was considered one of the most influential schools. In his oil-on-panel painting titled "Homage to the Square: Soft Spoken" (1969)[23] there are four almost concentric squares in different colours that provide an optical illusion about the painting's dimensionality. Art historians tend to include artworks of Albers both into the Op Art and the Post-Painterly Abstraction movements. Op Art, short for Optical Art, was the abstract art movement from the 1960s focused on exploring the capabilities of the human eye and its tendency to be erroneous. With vibrant colours, artists evoked optical illusions of dimensionality, movement, and shimmering of forms. Abstract artists, working in the twentieth century, abandoned the idea of art as imitation of nature, and painted forms that do not remind one of any specific objects. The paintings of the Russian/American Abstract Expressionist artist Mark Rothko (1903-1970)[24] present areas of glowing colours of considerable magnitude that provide the viewer with a colour experience invoking intellectual and spiritual connotations.

2.8.5.1 Primary colours in pigment and in light

Colours are important in the elements of design; the colour wheel is used as a tool, while colour theory provides a body of practical guidance for colour mixing and creating visual impacts of specific colour combinations.

The pigment colours are derived from mixtures of pigment primaries, the three primary hues which cannot be created by mixing. A pigment primary is caused by the reflection of two light primaries. The pigment primaries are red (magenta), yellow, and blue (a blue-green referred to as cyan). In practice, however, a more practical set of "double primaries" is utilized to allow for creating more intense saturation of colours.

All light colours are derived from mixtures of light primaries. A light primary is caused by the reflection of two pigment primaries. The light primaries are green, red-orange, and blue-violet. This means that a pigment primary is a secondary colour of light.

The value of colour is its lightness and darkness. Values[25], tints and shades[26] of colours can be created by adding black to a colour for a shade and white for a tint. Saturation[27] brightens a colour, makes the colour more vibrant than before. Creating a tint or shade of colour reduces saturation.

[23] http://www.metmuseum.org/toah/works-of-art/1972.40.7
[24] http://www.nga.gov/feature/rothko/classic1.shtm
[25] https://en.wikipedia.org/wiki/Lightness_(color)
[26] https://en.wikipedia.org/wiki/Tints_and_shades
[27] https://en.wikipedia.org/wiki/Saturation_(color_theory)

Colours can be made lighter or darker by adding either white or black. To lighten value, add white. Lightening (any colour plus white) produces a tint. Any colour plus black is a shade. Black plus white makes grey. A colour plus grey is called a tone. Black, white and grey are called neutrals. Hue, a synonym for colour, is a particular quality of a colour (full intensity, tint, tone, or shade). In order to change the hue of a colour, we add the neighbouring colour. Primary hues are red, yellow, blue. Secondary hues are orange, green, and violet. Intermediate hues are yellow-green, blue-green, etc. Intensity means the purity or strength (also called chroma). To change intensity and produce a tone, add a complementary or grey colour.

The colour star contains primary, secondary, and tertiary colours. Colour can add organisation when you develop a colour strategy. Colour can emphasise a hierarchy in an artwork. It is important too that colour choices in design change meaning within cultural contexts. For example, the colour white is associated with purity in some cultures while it is associated with death in others.

2.8.5.2 Project: Drawing an apple

Iconography – visual signs, symbols, and icons contained in the work – make for a rich set of meanings, connotations, and cultural traditions. In the case involving drawing an apple, and then writing a short poem about apple, it may serve as an example of translation from the visual to the verbal, and beyond the verbal. The choice of an apple as a theme may come from the fact that an apple is undeniably an iconic object – an image that represents some object that has a symbolic meaning beyond the object itself, filled with connotations, associations. It often serves as a metaphor, shown in educational or religious contents, or as a logo for Macintosh computers; it can even be found on a dentist's business card. Apple and apple pie have been deemed an American icon[28]. Steve Jobs knew well how many connotations it builds when he created his logo for the Apple Company [22]. Apple has been painted by masters representing almost all styles in painting – such as in Classical art, Impressionism, Mannerism, and Cubism. It has been presented in many ways: As a still life with apples on a plate, in a basket, as a scene with a woman eating an apple, a landscape with an orchard of apple trees, as an abstract, a cross section showing geometry of the seed placement, and in many other ways. An apple surely invokes connotations related to old scripts (such as biblical accounts), scientific research and experiments (Newton's apple), artworks (Claes Oldenburg), legends (Wilhelm Tell), symbols (an apple for a teacher), fairy tales (Snow White), artistic exaggeration (Giuseppe Arcimboldo), companies (Apple computers), places (Big Apple), many literary and cinematic works, and colloquial expressions (the apple of my eye, an apple a

[28] `http://en.wikipedia.org/wiki/Icon_(secular)`

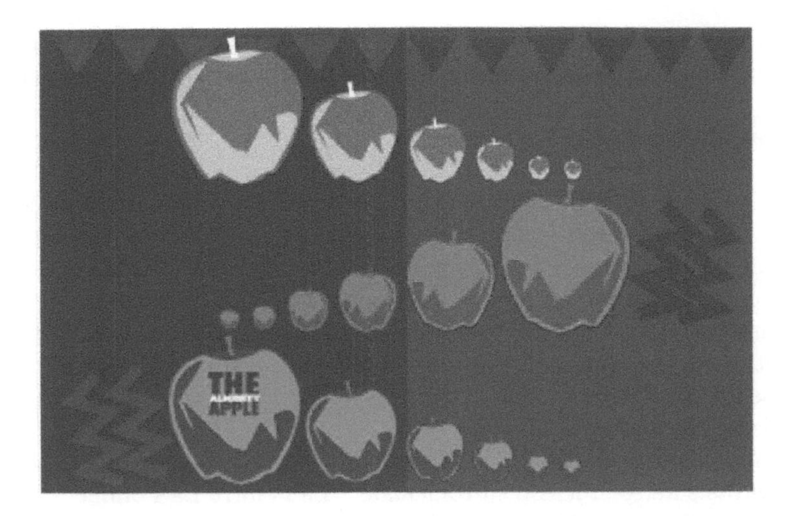

FIGURE 2.28 The mighty apple. © Matthew Rodriguez. Used with permission.

day keeps the doctor away, and so on). Figure 2.28 presents an image of an apple as a powerful icon in a computer graphics artwork. The coexistence of strong colours, dynamic composition, and the use of sharp shapes supports the message of the artwork.

Apple is related to Alan Turing, a British mathematical genius and a code breaker born 100 years ago. Turing died at the age of 41, allegedly because he committed suicide by biting an apple suffused with cyanide. However, at a conference held in Oxford in June 2012, Turing expert Prof. Jack Copeland questioned the evidence that was presented at the 1954 inquest; possibly, it was an accident related to chemical experiments, which Turing conducted in his house [38]. Until recently, Turing and his legacy were virtually unknown to the public because of active persecution caused by his homosexuality. Turing's Pilot Ace (Automatic Computing Engine) was faster than other contemporary British computers of the time by about a factor of five; it contained more technical detail than the American Edvac Report in dealing with software issues and in predicting future non-numeric applications of computers [30]. Using his computer, Turing broke the German wartime code Enigma that was used by the U-boats preying on the North Atlantic shipping convoys. Germany's encrypted messages were intercepted within an hour, sometimes less than 15 minutes after the Germans had transmitted them (total of 84,000 Enigma messages each month - two messages every minute). It was a massive code breaking operation. Breaking the U-boat Enigma code shortened the war in Europe by two to four years. At a conservative estimate, each year of the fighting in Europe brought on average about seven million deaths. The significance of Turing's contribution is in preventing the deaths of the further

14 to 21 million people that might have died if the U-boat Enigma code had not been broken and the war had continued for another two to three years [10].

Drawing an apple

Try drawing an apple such that it causes the same feelings invoked when you have a joyful, good feeling while looking at an object that really means something to you. Use it to convey this feeling to your friends by sending a simple drawing with a short text or posting it online. The first task is to focus on your own ability to look and see, for example, to convey your personal perception of the fruit and everything it could mean. The aim is to grasp the essence of an object, such as an apple, which was what Cezanne worked towards in his paintings mind[29].

Many years ago a French poet and screenwriter Jacques Prévert (1900-1977) wrote a verse "Picasso's Promenade" (Le Peintre La Pomme & Picasso, Paroles - Gallimard – 1949) – a collection of apple related connotations and addressed it to Pablo Picasso (1881-1973). The poem was set to music by Joseph Kosma and then sung by a prominent French vocalist Yves Montand in 1962. Below are short excerpts of this verse:

> On a very round plate of real porcelain
>
> an apple poses
>
> face to face with it
>
> a painter of reality
>
> vainly tries to paint
>
> the apple as it is
>
> but the apple won't allow it
>
> ...
>
> the apple disguises itself as a beautiful fruit in disguise
>
> and it's then
>
> that the painter of reality
>
> begins to realize
>
> that all the appearances of the apple are against him
>
> then suddenly finds himself the sad prey
>
> of a numberless crowd of associations of ideas

[29]For instance:

- `http://images.metmuseum.org/CRDImages/ep/original/DT1940.jpg` or
- `http://www.thegardenerseden.com/wp-content/uploads/2009/11/cezanne-apples1878-79-.jpg`

...

and the dazed painter loses sight of his model

and falls asleep

...

What an idea to paint an apple

says Picasso

and Picasso eats the apple

and the apple tells him Thanks.

Shape and Form

From the apple task, we see that all objects have shape or form. Shapes describe two-dimensional configurations. In an outline drawing we only show the shape. The shape of an object looks flat. Both shapes and forms are described by geometry. Geometric shapes such as triangles, squares, or circles have no volume; they are two-dimensional. Geometric forms have volume – a word we use when we describe the weight, density, and thickness or mass of an object.

Ambiguous shapes

An ambiguous shape is doubtful, uncertain, or open to more than one interpretation. A favourite textbook example of ambiguous space is a vase/profile, and an old/young woman. Another classic example of an ambiguous image is a drawing of a cube with lines and squares (see Figure 2.29). Our brain interprets the image as a three-dimensional cube. However, our brain is unsure where the closest part of the cube is. For many people, after looking for a while, the image seems to "jump" back and forth between two different cubes by visually switching their sides.

Form describes a three-dimensional object and gives the three-dimensional feel and look of an object. In drawing and painting, we can use shading and highlighting of an outline drawing to show this (Chiaroscuro). Forms have volume – a word that describes the weight, density, and thickness of an object. The solidness or volume of the form could be obtained by using highlights on one side of each object and shading on the opposite side. Shapes are geometric (such as triangles, squares, circles, etc.) or organic (such as leaves). Forms also are geometric (such as pyramids, cones, cubes, spheres, etc.) or organic – natural (like trees). They can be irregular (like clouds).

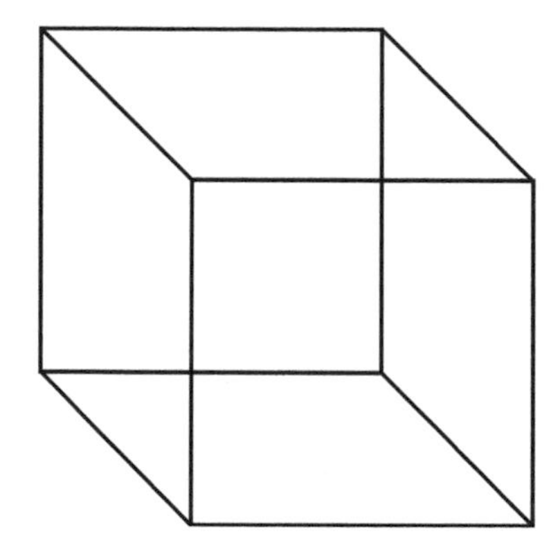

FIGURE 2.29 Cube.

2.8.6 Space

Space is the void between solid objects (forms) and shapes. It is everywhere, all around us. Everything takes up space in one form or another, whether it is two-dimensional, like drawing and painting, or three-dimensional, like sculpture and architecture. Paintings, drawings, and prints take up two-dimensional space. In a painting, it is limited to the edges of the canvas. Sculpture and architecture take up three-dimensional space. Music and literature involve time. Some arts, such as film, opera, dance and theatre take both space and time.

Positive and negative space

Space also describes the void between solid shapes and forms. The solid shape or form is called a positive space. The space within the drawn objects is a positive space; a doughnut has a positive shape or form. The space between the objects is a negative space; the doughnut hole is a negative shape or space.

Perspective

The depiction of depth or distance on a flat surface is called perspective. We live in a three-dimensional world with depth and distance, but we picture the world on a two-dimensional surface, such as on a piece of paper or a computer screen. The use of perspective enables us to show objects and scenes as they appear to the eye, with relation to implied depth on a flat surface of the picture. Designing a three-dimensional display may involve visual thinking in many ways. Perspective drawing is one of the solutions, with orthographic projection or the bird's eye view often applied. Other methods of preparing a display in three dimensions may include presenting the data on cardboard models, or using polygonal modelling, such as in a 3D program. Other presentation techniques include architectural miniatures, stereo illustrations and slides. Three-dimensional graphics are designed with the use of programming strategies or 3D graphics software packages; data can be presented as holograms, video, stereo illustrations, and slides. Maybe the most impressive way to show whatever you want is a computer-based immersive multi-wall virtual reality and interactive visualisation environment. Thus, we can show a 3-dimensional object as a 2-dimensional image when we draw a perspective drawing, a shaded drawing, or create a painting or a sculpture This is distinct from what we see on a map. We can consider a geometrical cardboard model a three-dimensional form of display, but we cannot say this about a musical score, a railroad schedule, a periodic system of elements in chemistry, or a tabular array of numbers (these are all two-dimensional).

Several methods help us create an aerial perspective, the representation of objects and scenes according to their distance form the viewer. Drafting a perspective includes overlapping, vertical positioning, greying colours, varying details, varying size, and converging lines. By overlapping objects we create the illusion that the partially hidden object is more distant. Vertical positioning lets us believe that objects placed higher are at a considerable distance and those positioned lower are closer in space. Gradation of the strength of light and colours of objects, and showing objects in greyed colours make them look remote, if we take into account the quality of light falling on the objects, and the surrounding atmosphere through which they are seen. We may show less detail on an object to make it appear more distant than another, as well as drawing distant objects smaller and nearer objects larger. Converging lines is the most frequently used method to demonstrate perspective – by drawing lines closer and closer together into the distance.

The Dutch graphic artist Maurits Cornelis Escher (1898-1972) explored a strange world of optical illusions, visual jokes, paradoxes, and puns. He drew ambiguous spaces to create illusions. In his work, often based on mathematical concepts, Escher combined humour, meticulous precision, logic, visual trickery, and beauty. Many of his works display metamorphosis, gradual transformation of one shape into another. Escher created impossible scenes by combining several kinds of distorted perspectives into a coherent whole. His work has been widely regarded as representative of mathematical thinking in art.

2.8.6.1 Showing space in a 2D drawing

Filippo Brunelleschi (1377-1446) first demonstrated his linear perspective during the Renaissance, the concept relating to how an object seems smaller the farther away it gets.

Before the invention of perspective, the following techniques for applying depth were utilised

(a) By overlapping objects, which appear to be on top of each other. This illusion makes the top element look closer to the observer, however one may determine the sequence of closeness but not the exact depth of space.

(b) By placing distant objects higher and closer objects lower

(c) By showing distant objects in greyed colours, and shading distant objects, which makes an object of a two-dimensional surface seem three-dimensional (Chiaroscuro).

(d) By showing less detail on distant objects

(e) By sizing: Showing distant objects smaller and closer objects larger

(f) By making lines come closer together in the distance

(g) By colouring the backplane darker or lighter than the front part of the scene.

Atmospheric perspective shows how air acts as a filter and changes the appearance of distant objects (the purple haze on mountains in the distance).

2.8.6.2 Perspective in painting

In his book entitled "Envisioning Information", Edward Tufte (1983; 1992) wrote that the world is complex, dynamic, multidimensional, and multivariate by nature, having many values and dimensions. By contrast, paper is static, flat. Hence, how to represent the rich visual world of experience and measurement on a flat piece of paper, Abbott's flatland? How can the artist escape flatland?

We can answer this question by examining how artists created perspective at different times. Florentine architects from the 15th-century Italian Renaissance perfected geometry and developed perspective drawing: For example, Fra Angelico "The Annunciation" (painted in 1441-3)[30], and also Sassetta, "The Journey of the Magi" (Siena, painted in 1435)[31]. Piero de la Francesca, "Portraits Federico da Montefeltro and Battista Sforza" (1465)[32]. Frantisek Kupka, "Cathedral" (1913)[33]. Vertical and diagonal geometric shapes are created in the pure abstraction style called Orphism, which was delivered from Cubism. Also, cubist artists, such as George Braque (1882-1963), Juan Gris (1887-1927), Pablo Picasso (1881-1973) showed a simultaneous view of many viewpoints. Marcel Duchamp "Nude Descending a Staircase" (1912)[34] was influenced by the geometric chronophotography of Marey. Etienne-Jules

[30] http://www.artbible.info/art/large/255.html

[31] http://www.museumsyndicate.com/item.php?item=23630

[32] http://www.wikipaintings.org/en/piero-della-francesca/portraits-federico-da-montefeltro-and-battista-sforza-1465

[33] http://en.wikipedia.org/wiki/František_Kupka

[34] http://en.wikipedia.org/wiki/Nude_Descending_a_Staircase,_No._2

Marey[35] and the 'cinematography' of Eadweard Muybridge (from the 1880s)[36] created more plastic cubism by representing a temporal fourth dimension on a two-dimensional canvas.

2.8.6.3 Cognitive perception addressed by painters

Artists' concepts related to photorealistic 3D rendering and to interactive visualisation were not always in perspective. Painters did not always use regular perspective or orthographic projections, with lines at right angles to the plane of the painting's surface. According to John Counsel [11], at times the Masters chose the kinds of perspective according to their personal cognitive perception. Sometimes, it was an isometric method of drawing, so that three dimensions were shown not in perspective, but in their actual size. For example, interiors painted by Vincent van Gogh were not fully based on the familiar Euclidean perspective. Cezanne's paintings reflected the way the viewer perceives reality. Areas where the viewers were supposed to direct and focus their eyes were painted in three dimensions, with greater detail and enhanced colour. The rest of the painting addressed peripheral vision and so it was flat.

Artists sometimes exaggerate parts of their paintings to stress the importance of the depicted objects. For example, in the style of Symbolism, in the mostly French artistic movement from the late nineteenth century, things were painted out of proportion at the cost of realism to convey otherworld ideas, unusual feelings, and states of mind. The giant's huge eye in "The Cyclops" created by the French painter Odilon Redon (1840-1916)[37] terrifies the viewer like in a nightmare. Artists painting in the Surrealist style created figurative portrayals of dreams mixed with reality aimed to liberate the unconsciousness. An acrylic painting on panel, "Death and Funeral of Cain" by David Alfaro Siqueiros (1896-1974)[38], influenced both by the Surrealists and the Mexican folk art, shows an enormous, out-of-proportion dead chicken.

2.8.7 The principles of design in art

Principles of design that are most often used in the visual arts are balance, emphasis, movement, variety, proportion and unity. These principles vary according to the artist using them. For example, some textbooks discuss also contrast, rhythm, and repetition [16]. They answer the question of how we use the elements of design. When applied to the *elements of design, principles of design* bring them together into one design and determine how successful a design can be.

---

[35] http://stage.itp.nyu.edu/history/timeline/marey.html

[36] https://commons.wikimedia.org/wiki/File:Eadweard_Muybridge_Gehender_Strau%C3%9F_001.jpg

[37] http://en.wikipedia.org/wiki/File:Redon.cyclops.jpg

[38] http://www.terminartors.com/artworkprofile/Siqueiros_David_Alfaro-Death_and_Funeral_of_Cain

2.8.7.1 Balance

Balance is about the arrangement of lines, colours, values, textures, forms, and space, so that one section or side of an artwork does not look heavier or stronger than another. We can see in art three main types of balance: Formal (or symmetrical), informal (or asymmetrical), and radial balance. Formal or symmetrical balance has equal weight on both sides (for example, two people of the same weight at the two ends of a seesaw). Informal or asymmetrical balance has a different weight placed differently on each side to maintain balance (for example, when a heavy person on a seesaw sits closer to the centre and the lighter person sits farther out on the end). Radial balance is a circular balance moving out from a centre of an object to maintain balance (for example, when only one object is centred in a picture). Many times, we see a combination of formal and radial balance, or the informal balance with the unequal organization of elements; for example, several small shapes may juxtapose a large shape.

We may want to examine how artists create balance. For example, Anish Kapoor[39] achieved in Double Mirror, 1998[40] balance in three dimensions of his illusionary void, while Edward Hopper, People in the Sun, 1960[41] in a painting, with an informal balance that shows the stillness of the scene with strongly contrasting lights and shadows.

2.8.7.2 Emphasis

Emphasis is a way of bringing dominance or subordination into a design or a painting. Major objects, shapes, or colours may dominate a picture by taking up more space (when they are larger and repeated more often), by being heavier in volume, or by being stronger in colour and colour contrast than the subordinate objects, shapes, or colours. There must be a balanced relationship between the dominant and subordinate elements; otherwise there is too much emphasis. Colour and colour contrast can be used to achieve emphasis in a work. Also texture contributes to emphasis.

2.8.7.3 Movement

The illusion of movement may be achieved by the use of lines, colours, values, textures, forms and space, to direct the eye of the viewer from one part of the picture to another. For example, the feeling of movement can be suggested by an arrangement of shapes. Some possibilities are circular, diagonal or vertical arrangements; we may also create illusion of movement that goes back into the distance, both by diminution of sizes and similarity of shapes.

---

[39] https://www.theartstory.org/artist-kapoor-anish.htm

[40] http://art-glossary.com/definition/anish-kapoor/

[41] http://www.wikipaintings.org/en/edward-hopper/people-in-the-sun

We may want to look at how artists employ some elements of art such as line or space, and apply emphasis, rhythm, or contrast to enhance the dynamics of their artwork and show motion in various styles in art such as: Symbolism. For example, in "Centaurs' Combat"[42] the Swiss painter Arnold Böcklin (1827-1901) depicted a mythological scene of combat to convey the savage and violent nature of ferocious fight. There are the contrasts in colours, contorted shapes, and unreal representation of the background space, all of them enhancing the sense of horror and drama. In Cubism, for example, Fernand Léger (1881-1955) "The Builders"[43]: Léger used horizontal and vertical lines defining the space of the ironwork skeleton. At the same time, he used colours and patterns to emphasise the dynamics of the scene. Artists belonging to the Futurism movement were intrigued with motion through time and space. Examples inlcude: Umberto Boccioni (1882-1916)[44], 1911[45], Giacomo Balla (1871-1958), "Flight of the Swallows"[46]. In "Flight of the Swallows" the artist used repetition to enhance the feeling of movement. Works done in the Surrealism style, for example, Arshile Gorky (1904-1948), "The Waterfall"[47], is almost abstract but conveys a strong impression of water pouring through a rock. In the semi-abstract art of Wassily Kandinsky (1866-1944), for example, "Cossacks"[48], his colourful brush strokes, dynamic lines and shapes, create a wonderfully balanced composition [26]. And, in dynamic paintings of the American painter Thomas Eakins (1844-1916), for example, his scenes of wrestling[49].

2.8.7.4 Composition

In art, composition is the orderly arrangement of visual elements, a proper combination of distinct parts so they are presented as a unified whole. The whole arrangement of objects in a picture is made to the best advantage of its elements and defines how everything is put together using thoughtful choices. To develop composition in a drawing or painting, we have to select the objects we want to show, then create a centre of interest and find balance among the objects. Good composition may involve movement, rhythm, and well-arranged positive and negative space. In graphic design and desktop publishing[50] composition determines the page layout.

It took many years to develop routine features of books, with page numbering, indexes, tables of contents, and title pages. Web documents underwent

---

[42] http://www.wikipaintings.org/en/Tag/Centaurs#supersized-search-260594
[43] http://www.amazon.com/Hand-Made-Oil-Reproduction-Builders/dp/B004LA7Q8E
[44] http://en.wikipedia.org/wiki/The_Street_Enters_the_House
[45] http://en.wikipedia.org/wiki/Umberto_Boccioni
[46] http://www.wikipaintings.org/en/giacomo-balla/flight-of-the-swallows-1913
[47] http://www.wikipaintings.org/en/arshile-gorky/waterfall-1
[48] http://www.wassilykandinsky.net/work-250.php
[49] http://commons.wikimedia.org/wiki/File:Thomas_Eakins_-_The_Wrestlers_(1899).jpg
[50] http://en.wikipedia.org/wiki/Graphic_design

a similar evolution and standardisation in order to define the way information is organised and made available in electronic form. Visual composition of a website and its graphics is an important part of the user's experience. In interactive documents, the interface design includes the metaphors, images, and concepts used to convey function and meaning of the website on a computer screen.

In accordance with Edward Tufte [49], every graphic presentation, as well as every project should fulfil general principles related to its design, and should pass critical evaluation of its editing, analysis and critique for presentation. Edward Tufte argues that information should enhance complexity, dimensionality, density, and the beauty of communication [49, 48]. Good information display should be: Documentary, comparative, casual and explanatory, quantified, multivariate, exploratory, and sceptical; it should allow comparing and contrasting. When envisioning statistical information, such displays should insistently enforce comparisons, express mechanisms of cause and effect quantitatively, recognise the multivariate nature of analytic problems, inspect and evaluate alternative explanations.

We may want to examine the composition of some artworks. For example, the sculpture by Constantin Brancusi "Bird in Space" (1911)[51] has the most basic, abstract primordial vitality. And, in Pierre Bonnard "The Dining Room in the Country" (1913)[52], the composition of both the outdoor and indoor scene has been achieved by using broad areas of colour.

2.8.7.5 Variety and contrast

An artist uses elements of design to create diversity and differences in their artwork. Contrasting colours, textures, and patterns all add interest to the artwork. Highlights of colour to the corners or edges of some shapes may be used to add contrast. To examine variety and contrast in some artworks, we can see how the Dutch graphic artist M. C. Escher (1898-1972) used contrast to isolate some shapes from others and from a main body of his works[53], or Wassily Kandinsky, "On White II" (1923)[54]. Kandinsky's analytical book, "On the Spiritual in Art" (Kandinsky, 2011) was first published in December 1911. Kandinsky was one of the co-founders of The Blaue Reiter group. His work was shown in New York in 1913 at the Armory Show. He taught at the Bauhaus school of modern design. Born in Russia in 1866, Kandinsky became a German citizen in 1928. The Nazi government closed the Bauhaus in 1933 and later that year, Kandinsky settled in Neuilly-sur-Seine, near Paris; he acquired French citizenship in 1939. The Nazis in the 1937 purge of "degenerate art"

[51] http://en.wikipedia.org/wiki/File:Bird_in_Space.jpg

[52] http://en.wikipedia.org/wiki/File:Bonnard-the_dining_room_in_the_country.jpg

[53] http://en.wikipedia.org/wiki/File:Escher_Circle_Limit_III.jpg

[54] http://en.wikipedia.org/wiki/File:Kandinsky_white.jpg

confiscated fifty-seven of his works. Kandinsky died December 13, 1944, in Neuilly.

2.8.7.6 Proportion

The size of one part of an artwork in comparative relation to its other parts is called proportion. Artists use proportion to show balance, emphasis, distance, and the use of space. Sometimes, proportion is used to add emphasis to an artwork. For example, in medieval religious paintings, some rulers or saints are pictured out-of-proportion to emphasize their importance: Important figures were painted bigger and humble donators smaller.

We may want to examine proportion in some artworks. For example, in Cimabue, "The Santa Trinita Madonna" (c.1280) we can see small figures at the bottom[55]. In Jasper Johns' "Three Flags" (1958)[56], flags of different sizes were superimposed on top of one another. We can see the use of proportion to evoke the feeling of dimensionality and elicit some optical effects characteristic of the American Pop Art style.

2.8.7.7 Unity/harmony

Unity is the result of how all the elements and principles of design work together. All parts must have some relation to each other. They must fit together to create the overall message and effect. According to the author of *The Elements of Graphic Design*, Alex White [53], to achieve visual unity is the main goal of graphic design. When all elements are in agreement, a design is considered unified. No individual part is viewed as more important than the whole design.

2.8.7.8 Analysis of artworks

In the oil on canvas painting "Saint Francis of Assisi" created by the Spanish artist Francisco de Zurbaran (1598-1664), the expression of strong religious feelings is enhanced by both the ecstatic look on the saint's face and the dramatic effect of the *chiaroscuro* lightning, strong from the left side only. The artist chose the dark colour of the background and the elongated form of the figure to place emphasis on an austere religiousness conveyed in the artwork. "The Treachery of Images" (1928-9)[57] by Belgian artist Rene Magritte (1898-1967) brings a challenge to the ordered society by denying that a picture that is obviously an image of a pipe is a real pipe. A pipe has been pictured in a clear and sharp way expressing an awareness of an object as it really is. Thus, the artist declares, in accordance with the Surrealist movement in art, that all is

[55] http://en.wikipedia.org/wiki/File:Cimabue_033.jpg
[56] http://en.wikipedia.org/wiki/File:Three_Flags.jpg
[57] http://en.wikipedia.org/wiki/File:MagrittePipe.jpg

not as it appears to be, and an image of an object cannot be confused with the tangible reality. Pablo Picasso (1881-1973) created a series of paintings titled "Weeping Women", along with his "Guernica," telling about the cruelty and violence imposed by the Spanish Civil War in 1936. The face of the "Weeping Woman" (1937)[58] is shattered in the style of the Cubist movement. Strong colours, firm paint strokes, and sharp lines create the dramatic expression of the weeping woman. The same concern has been imparted quite differently by Robert Motherwell (1915-1991) in his acrylic on canvas painting "Elegy to the Spanish Republic No. 134" (1961)[59]. Strong, dynamic shapes are painted in black on a white background in the style of the Abstract Expressionist movement.

2.8.8 Basic art concepts

Many concepts about art stem from the generally accepted opinion that every artwork, as well as every project, should fulfil general principles related to its design, editing, and analysis of presentation. We will talk more about these qualities, especially beauty and aesthetics of display. For now, in order to better analyse and appreciate a work of art, it may be useful to focus on basic art concepts that can apply to any work of visual art. These basic art concepts include the type or form of art, subject matter, style, medium, and design.

If we create an artwork representing an apple, we need to make certain decisions (after acquiring an apple):

(1) **Type or form of art** describes the kind of art or art products, the schemes used to classify art, and the functions art products serve; an artwork may be described as a drawing, a painting, a sculpture, etc. Would we paint, sculpt, draw an apple, or erect an architectural structure in the form of an apple?

(2) **Subject matter** defines the meaning of the work of art; the theme, topic, or motif represented as a person or object; it may be a portrait, a landscape, a still life, an abstract work, etc. Would we represent an apple as a part of a still life, a portrait of a man or a woman eating an apple, a landscape such as an orchard with small, multiplied images of an apple, an abstract artwork about an apple, or an almost abstract image of a close-up of the apple skin or showing an apple's interior?

(3) **Style** tells us about the traits and resemblances within a group of works of art, the visual similarities influenced by the time, place, or personal manner of the artist. We need to decide if an apple would be seen as a geometric form, as the cubist artists would see it; if we work on the scenery issues, like in

[58] `http://picassogallery.blogspot.com/2011_01_18_archive.html`
[59] `http://www.metmuseum.org/Collections/search-the-collections/210009638`

paintings from the Renaissance or Mannerism periods; if we focus on light and paint related work, as was done by the Impressionist artists; or if we create a generative artwork using programming or software.

(4) **Medium** informs us about the materials, tools and procedures the artist uses to create the work of art. For example, medium can be described as: 'Oil on canvas mounted on panel,' 'acrylic on canvas,' 'marble sculpture,' 'paper, pen and ink over chalk drawing,' or 'polychrome woodblock print on paper.' We need to secure proper materials, tools, and apply selected techniques for our representation of an apple. The medium can depict the differences resulting from a particular use of materials and tools. For example, an oil-on-canvas painting will look different from an oil-on-paper painting because a canvas repels paint, while paper absorbs it. That is why we refer to the artwork as 'oil-on-canvas painting' rather than just 'oil painting.'

(5) **Design** specifies the elements and principles used in the artwork; the planned organisation of the visual phenomena the artist manipulates. We can cut an apple in half and focus on pips (seeds) and how they are placed inside five carpels arranged in a five-point star. We can picture the repetition of lines and dots on the apple skin or exaggerate irregularities in its design.

(6) **Interactivity** in cases of creating time-based works, we need to examine how the users would interact with the entire setup.

From these descriptions of the elements of design hopefully you can see how they carry over to design using computers and programming. We will discuss this transition in more detail in the next section.

2.8.9 Visual aspects of mathematics, computing, and programming

Scientists and artists see the purpose in applying visual ways of presentation while working on particular scientific branches. For example, mathematicians, anthropologists, artists, and designers; architects conduct computer analysis of facades, friezes, and some architectural details. Researchers in many fields of natural sciences, medicine, pharmacology, biology, geology, or chemistry, examine and visualise symmetry in natural and human-made structures. And, many artists have created masterpieces this way.

Visual aspects of mathematics and computing allow presenting theories and their proofs as two dimensional and three dimensional constructs. Placing emphasis on developing visual literacy in students may be beneficial in fulfilling this task. For example, students may create mathematically programmed sculptures, or fractal based artworks, sceneries, and backdrops. Some versed in programming could be coding music- and dance-related shapes and form, thus creating music visualisations. Current means of delivering knowledge, for example, with the use of 3D printing technologies [34], augmented reality [7], and open source printers are beginning to appear in the school environment [21, 45]. In 3D printing, based on the rapid prototyping process, additive processes are used, with successive layers of thermoplastic material laid down according to a computer program [18, 41]. Thus, for example, the Voronoi tower can be created. Two- and three-dimensional Voronoi diagrams (Voronoi tessellations) are now commonly applied in architectural concepts, and other science and technology fields.

Artists transform patterns and repetitions to apply the unity or symmetry in their compositions (for example, by examining a Fibonacci sequence, prime numbers and magic squares, a golden ratio, and tessellation techniques). Mathematicians, computing scientists, and artists apply visual metaphors as a cognitive tool to visualise the world's structure and our knowledge. For example, hierarchical structures are predominantly analysed with the use of a tree metaphor. Manuel Lima called the tree figure the most ubiquitous and long-lasting visual metaphor, "through which we can observe the evolution of human consciousness, ideology, culture, and society" [31, p. 42].

CHAPTER 3

Coding for Art

CONTENTS

3.1 INTRODUCTION

THERE are many different types of art practised. In the visual arts (paintings, drawings, etc.) there are four main categories: The portrait and group portrait; landscape; still life; and abstract. However, many more categories have been added since the advent of computers. For example, interactive art was first developed in the 1960s by Myron W. Kruger, who also worked as an early researcher of virtual reality and augmented reality art. Advances in computing supported progress in installation art, performances, mathematical art, bio-inspired art, nano art, visualisation based art, visual music, simulations, robotics based art, 3D printing, wearable art and works based on ubiquitous apps, new materials, fibre optics, virtual reality (VR), and more. The advanced section of this book is addressed to those who want to experiment beyond the traditional art categories.

In this part of the book, the user, as a new programmer, can create graphics according to four types of topics: Portrait/group/self-portrait, landscape, still life, and abstract (for example: A Close-Up, Micro/Macro, or Nano, or geometry-based). As a coder you can also organise the composition, focusing on selected elements and principles of design. This is a merging of knowledge beyond coding theories, languages, around the basics of art theories. Some design and craft strategies will be applied. Most assignments could go from the static image to animation [5].

3.2 CODE ART

In this section, a few projects along with their relevant code are presented. First, portrait and still life are transformed into ASCII art, and then, a project showing the process of illustrating a scene of a cherry tree is presented; this is then followed by another project to create a map of winds, and then students' reaction to an interactive website related to wind mapping is provided.

3.2.1 ASCII art

By Alireza Ebrahimi
SUNY, Old Westburry, Long Island, New York

3.2.1.1 A personal account

Early computers only dealt with formulas. Early machines were number crunchers and used for numerical problem solving; anything else was outside the box of what the computer was designed for. The following is a personal account of how I used basic computer art.

As a student of computer science, in 1977, I used special characters, such as asterisk '*', dash '-', underscore '_', plus '+', and other symbols, in the BASIC programming language, with a simple series of print statements, in order to draw geometric shapes, flowers, snowmen, bunnies, and even Snoopy. Later on, creating calendars for the years to come, and even past years, was exciting. As a beginner programmer this sort of "artwork" could be put together using a series of print statements. The if statements, loops, and functions eliminated redundancies in code and made the task progressive.

It was rewarding to view results in early interactive computers using monitor terminals by Xerox Data Systems, Sigma 9. You got to the point that you could draw anything, and the first task was to draw it manually and map it by using any special characters so that it would resemble the actual drawing work. For example, placing dashes next to each other could be used to make a line. Despite the fact that punch cards (80 columns wide) were still common in programming, the punch cards and card readers made it a tedious task to do anything creative, as making one simple mistake (punching a hole) could lead to losing cards, which was also a financial burden. The output would be in the monitor room in the users' bin folder displayed on rolling papers (120 columns wide). Plotters were an output peripheral for most of the scientific and professional display of programs and were used as well for Computer-Aided Design/Computer-Aided Manufacturing (CAD/CAM). It gave a hard copy of the output and used a pen to draw machine images and architectural graphics.

Later, introducing fractal programming created another dimension to computer art by generating amazing image patterns, and allowed certain aspects of images to be repetitive and to be placed in different locations. I learned that I could draw two ways: 1) by displaying the special characters next to each other 2) finding a formula and mapping the x and y coordinates. Some images that could be generated included hearts, mountains, and flowers. Animation was the next step in the progression of computer art, by redrawing the frame in the next location using coordinates.

The experience of a program which prints a page of space on continuous rolling green bar paper on a line printer landed me a job as a computer aide working in a computer lab with a wage of $2.50 per hour, when pizzas cost 50 cents! Ironically, my school friends worked for a Persian Rug Business for $25.00 per hour. I studied in a computer lab, and since it was quiet for me, it was an alternative place to the library and I was able to find scratch paper for my homework, from the paper disposal from printing errors by other students learning programming. One day, when there was no scratch paper, I decided to take a page of paper from the roll; however, the lab manager yelled that the printers and papers are only for programmers! So I learned to program a print statement with a loop of 25 times in order to get a blank page. Then, when I needed two pages, I programmed with a counter that never reached the number 2. The initial number was 1; however, the increment was accidentally set to 2 so the counter went from 1, 3, 5, etc. This led to the only printer in the lab repeatedly printing blank pages and causing noisiness. The manager worried but had no idea what to do and what caused the problem. Then, I removed my program and fixed the problem. Because of this, the manager offered me work the next day. That simple artwork of displaying an infinite line of blanks helped me get my first job in the computing field!

Understanding the displayable character code and assigned numerical values of a particular computer system was a key to drawing artwork. The catch was that every character on the keyboard had a unique numerical assigned value. Programming in the CDC mainframe computer, there was a mysterious problem that I needed to resolve. There was a list of names that were sorted alphabetically; however, the last one was not in order. When I was consulted, I instructed them with humor to remove the last name in the list. They followed this and the list was now all in sorted order. But, the goal was to include all of the names. I instructed them again to add the name to the input file and run the program. This time the sorting was perfect. The prophecy was that the last name in the list was proceeded by a blank giving it a high value. The character value for the blank was 45 and letter 'A' – 'Z' was from 1-26, respectively. Interestingly, I found out that in early CDC code the character 0 had the decimal value of 27, and colon ':' (the first value) had the value 0, while the semicolon ';' (the last value) had the value of 63. I learned it the hard way; that my sorting list program showed a name with letter Z on the top. After I deleted the name and retyped it, I realised the sort was correct. The mystery of the story was that there was a blank space before the name, and a blank space before a letter z for a name will place it on the top of an ascending sort instead of being the last name.

Character codes were represented differently by different companies before 1963. As a way to unify the code, ASCII took the lead. ASCII (pronounced as AS-KEY) stands for a committee that established a code known as the American Standard Code for Information Interchange. ASCII is the continuation of what was originally telegraph code by the Bell company. ASCII, when originally released in 1963, represented every character on the keyboard and more. Later with 8 bits in 1979, known as "extended ASCII", it represented graphical characters for drawing, such as boxes and shades, bringing more life and creativity to ASCII art. These are some of the decimal values that a beginner programmer might need to know:

- Upper 'A' = 65, Upper 'Z' = 91 (65+26),
- Lower 'a' = 97, Lower 'z' = 123 (97+26),
- Blank ' ' = 32,
- Character '0' = 48,
- Character '9' = 57 (48+9).
- Character null = 0.

The ringing bell in ASCII is code 7, which can be programmed to sound the bell, beep, and even make a simple song like Jingle Bells.

The following demonstrates how to display ASCII art and use the C++ programming language. In order to use a high level language like C/C++, we need a compiler to translate the high level commands to its equivalent low level machine-understandable language. Depending on your operating system, there are different compilers available. Here are some suggestions: gcc for Linux, clang for OSX, and TDM gcc for Windows or Visual C.

Once the installation is complete, you can compile the following code, which returns "*Hello World!*" on your screen:

Listing 3.1 Hello World in C++

```
1 #include <iostream>
2 using namespace std;
3
4 int main() {
5   std::cout << "Hello World!" <<std::endl;
6   return 0;
7 }
```

FIGURE 3.1 Protrait of Ada Lovelace. © Alireza Ebrahimi. Used with permission.

And the next code uses "\n" to create a new line:

Listing 3.2 Separating lines in C++

```
1  #include <iostream>
2  using namespace std;
3
4  int main() {
5    std::cout << "Hello World! \n Welcome to the first C++ code" <<std::endl;
6    return 0;
7  }
```

And therefore displays the following (two sentences, each in one separate line):

```
Hello World!
Welcome to the first C++ code
```

3.2.1.2 Portrait and still life as ASCII art

Now that we have seen how to display some characters, we can look at how to convert an image into characters and display them. To convert an image to printable ASCII, there are a variety algorithms. Each generates a different type of ASCII art as output. There is an algorithm that converts a bitmap image to solid greyscale ASCII art. Each pixel from the original image maps to a certain character. However there are several free Image to ASCII art generators that can do this job for you. But if you feel brave, you can write your own algorithm which takes an image and returns the "equivalent ASCII characters.

FIGURE 3.2 A still life: An apple, a Persian rug, and a book. © Alireza Ebrahimi. Used with permission.

ASCII characters can be used to represent a portrait image by uploading an image through an ASCII art generator web site to produce the ASCII art[1]. Figure 3.1 shows Ada Lovelace and her ASCII representation. Ada Lovelace, an English mathematician and writer, and daughter of Lord Byron, was the first programmer and was engaged with the Analytical Engine of Charles Babbage in 1833. In another sample output, see Figure 3.2.

In order to generate ASCII art, pick a photo of your choice, convert it to ASCII characters (either by using an online image to ASCII converter, or by writing your own algorithm), then display the output in the command line by replacing the "Hello World!" with the ASCII characters; remembering to include `\n` at the end of each line, or search for an alternative method where the characters are displayed as they are pasted in the code.

The next part touches on the concepts of landscape which represent our outdoor surroundings. It varies from the flatness of an ocean to raised mountain peaks or volcanoes. It can contain fauna, flora, or people with endless combinations.

3.2.2 Landscape: A cherry tree

By Matt Anderson
Student at the University of Northern Colorado

For this project, I used Adobe Animate (formerly Adobe Flash) along with its new CreateJS Toolkit to create a simple scene of a cherry tree. Clicking start triggers a simple animation of leaves falling from the tree. Clicking wind triggers a (very rudimentary) rustling effect within the tree itself. Clicking

[1]`https://www.text-image.com/convert/ascii.html`

sunset transitions the scene to night time, with a simple setting sun and colour shift effect.

3.2.2.1 How does it work?

Using the CreateJS Extension and HTML5 Canvas in Animate allows you to assign JavaScript code to specific keyframes of an animation. CreateJS has Javascript methods that allow you to pause, play, and jump to specific keyframes. Used in combination with event handlers like buttons, a simple animation such as this can be created with a few lines of code.

3.2.2.2 The code

```
1  this.stop(); //Prevent the animation from playing automatically
2
3  this.playBtn.addEventListener("click", playClicked.bind(this));  //Add
       event listeners to all three buttons
4  this.windBtn.addEventListener("click", windClicked.bind(this));
5  this.sunsetBtn.addEventListener("click", sunsetClicked.bind(this));
6
7  function playClicked(){  //Functions called by event listeners
8    this.gotoAndPlay("beginPoint");
9  }
10 function windClicked(){
11   this.gotoAndPlay("loopPoint");
12 }
13 function sunsetClicked(){
14   this.gotoAndPlay("sunPoint");
15 }
```

`this.stop();` pauses the animation on the current frame. In this case, the JavaScript is contained within the first keyframe of the animation.

`this.playBtn.addEventListener("click",playClicked.bind(this));` is an example of an event listener. The code says: When the object called `playBtn` is clicked, call the function called `playClicked()`. The method `addEventListener` takes two arguments, in this case "click" and `playClicked.bind(this)`. The first dictates what type of event the listener should wait for, the second determines what should be done when the event is triggered.

```
1 function playClicked(){
2   this.gotoAndPlay("beginPoint");
3 }
```

This is the function that is called when the `playBtn` is clicked. It moves the current state of the animation to a specific point, "`beginPoint`" and automatically plays. In Adobe Animate, a label can be assigned to a keyframe to be later referenced through JavaScript. In this case, I created a label on the first keyframe called "`beginPoint`".

3.2.2.3 *What is 'this'?*

In these examples, you will notice I have used the word `this` before many method calls. In Object Oriented Programming, `this`, `self`, or `me`, dictates that the instance of the object being modified exists within the current scope of the program or project. In Adobe Animate, this refers to the entire scope of the project – more specifically, *this* project. The full code is provided in the book website.

3.2.3 Project: Map of winds

By Matt Anderson
Student at the University of Northern Colorado

Below is an attempt to create a map of winds, using Processing.

Listing 3.3 Map of winds

```
1  PulseRay rayOne, rayTwo, rayThree, rayFour, rayFive, raySix, raySeven;
2  WindBlower WB;
3  PImage windBlower, paperTexture, sun;
4
5  void setup(){
6    size(640, 360, P3D);
7    stroke(240);
8    windBlower = loadImage("wind.png");
9    paperTexture = loadImage("paperTexture.jpg");
10   sun = loadImage("sun_vector.png");
11
12   //Initiate Rays
13   rayOne = new PulseRay();
14   rayTwo = new PulseRay(30);
15   rayThree = new PulseRay(60);
16   rayFour = new PulseRay(75);
17   rayFive = new PulseRay(15);
18   raySix = new PulseRay(345);
19   raySeven = new PulseRay(115);
20
21   WB = new WindBlower();
22   smooth(6);
23
24   ps = new ParticleSystem(new PVector(10, 0));
25   pOne = loadImage("particle_1.png");
26   pTwo = loadImage("particle_2.png");
27   pThree = loadImage("particle_3.png");
28   pFour = loadImage("particle_4.png");
29 }
30
31 void draw(){
32   image(paperTexture, 0,0);
33   noTint();
34   tint(255,100);
35   image(sun, width/2 - 75, height/2 - 75, 150,150);
36   //Wind Blower
37   WB.display();
38
39   //Draw Rays
40   rayOne.displayRay();
41   rayTwo.displayRay();
42   rayThree.displayRay();
43   rayFour.displayRay();
```

```
44     rayFive.displayRay();
45     raySix.displayRay();
46     raySeven.displayRay();
47
48     //Generate Particles
49     pushMatrix();
50     translate(25,25);
51     rotate(radians(30));
52     ps.addParticle();
53     ps.run();
54     popMatrix();
55
56   }
57
58
59   class PulseRay{
60     float angle;
61     float startX;
62     float startY;
63     float xoff;
64
65    PulseRay(float angle){
66      this.angle = angle;
67      xoff = random(0,3.5);
68    }
69    PulseRay(){
70      this.angle = 45;
71      xoff = random(0,3.5);
72    }
73
74     void displayRay(){
75     pushMatrix();
76     translate(75, 75);
77     rotate(radians(angle));
78     line(0, 0, perlinNoise(), 0);
79     popMatrix();
80    }
81
82    float perlinNoise(){
83     xoff += .01;
84     float xNoise = noise(xoff) * width/3;
85     return xNoise;
86    }
87   }
88
89   class WindBlower{
90     void display(){
91       pushMatrix();
92       translate(50,-25, 1);
93       rotate(radians(45));
94       tint(255, 126);
95       image(windBlower, 0,0, 100,100);
96       popMatrix();
97     }
98   }
```

See referenced images in Figure 3.3.

3.2.4 Wind visualisation

Below are students' reactions to the interactive website "Wind Map", `hint.fm/wind`, showing data visualisation. This website may serve as an inspiration for readers' coding projects.

FIGURE 3.3 Map of winds: Referenced images.

The wind map is an intriguing, artistic study of natural elements. It is an example of how data can be represented in a visual, interactive form. Data concerning wind directions and speeds is collected from locations across the country. Instead of placing the data in a chart or table, it is mapped visually so that recognisable patterns of various places can be seen: Patterns of where the wind is, which direction it is blowing, how fast it is blowing, and possible implications, consequences, or results of converging or diverging wind patterns.

A large number of sensors are used to connect large data to a computer program to produce such up-to-date images. It is almost beyond comprehension the amount of sensors it takes to be connected to a computer program to produce such up-to-date images. The data comes from the National Digital Forecast Database. The wind patterns and forecasts are sourced once an hour, leaving the map of the United States as a living portrait. This project blends modern science and technology with creativity, incorporating both science and art in one project.

Presenting the data in a visual format makes it more accessible to most people. If this data were presented by simply listing the measurements for various points, it would most likely only be understood and interpreted by a scientist trained to read data in this manner. For those without a background in science, this collection of locations and numbers would not mean much. However, when this data is structured in the form of an interactive map, it is easier to recognise various places and patterns.

The interactive wind map shows the speed and direction of wind movement over the US at that time. It shows the current wind currents throughout the United States starting at 1 mile per hour (mph) and reaching 30 mph. The closer one zooms in, the more specific the wind speed gets. Also, the closer you zoom in, the more cities appear, allowing you to navigate through the US. The map is not completely accurate because it is more a piece of art than a scientific navigation tool. The wind map is interesting because it uses a naturally occurring system and translates it into art. The website has a gallery of screenshots that include days of hurricanes where the wind patterns are absurd or days that are very windy. This type of technology might be good to use in relation to traffic patterns on the road, train, planes, etc.

An interesting thing about the wind map is the circular motions and curved patterns that follow and repeat themselves. The map calculates the total wind and average wind in miles per hour for each day. The interactive wind map uses a naturally occurring system and translates it into art. The artist explains that this is their own personal art project and does not reflect an association with any company. This website also includes snapshots of previous hurricanes and strong winds that have affected parts of the United States, such as Hurricane Sandy and Hurricane Isaac.

Beneath the wind map the artist wrote, "An invisible, ancient source of energy surrounds us – energy that powered the first explorations of the world, and that may be a key to the future."

It looks a bit like a Van Gogh painting, in the ways the lines are close together or far apart depending on the wind speeds. Like in a Van Gogh painting, the lines are thin and create patterns in the lines between.

3.3 ABSTRACT ART

3.3.1 Introduction to geometric art with Python

By Stuart Smith
University of Massachusetts Lowell

Below, is an example of an abstract project, showing the use of elements and principles of design. It includes code developed by Stuart Smith, professor (emeritus) in the departments of Music and Computer Science, at the University of Massachusetts Lowell, for an abstract geometric composition using Python along with an introduction on how to set up this environment for coding:

This section shows how intricate geometric patterns can be created with relatively small, simple programs written in Python. Python is a general-purpose programming language that is currently used for almost every kind of computer application. Python is an *interpreted* language, which means there is no compilation step. You just type in commands and Python immediately executes them. Of course, you can also build and save programs for later execution. In this section you will use Python's built-in "turtle" graphics capability to make a specific family of geometric figures.

3.3.1.1 Downloading and installation

The first thing you need to do is download the current release of Python, which is free for non-commercial use. For step-by-step download and installation instructions for Windows, go to
`https://www.python.org/downloads/windows/`

For Macs, go to
`https://www.python.org/downloads/mac-osx/`

For Linux, Python comes installed already with the operating system.

After installing Python, you'll find it worthwhile to go here for a collection of classic Python demos:
`https://code.google.com/archive/p/python-turtle-demo/downloads`
Download and extract *Python turtle graphics demo suite for Python 3.1*. Each folder contains several demos. Double click on whichever one you want to see (Try several as some may not work in Windows).

3.3.1.2 Launching Python and entering commands

Once installed, Python is launched from the Start window. Click on Python 3.6, and in the dropdown menu click on IDLE (Python 3.6). The Python command window will appear. The >>> prompt on the left edge of the window indicates that Python is waiting for you to enter a command. You are now ready to make your first image. Just enter each of the following commands at the >>> prompt and watch what happens.

```
>>> import turtle
>>> tom=turtle.Pen()
>>> tom.pensize(3)
>>> tom.color("red")
>>> tom.circle(100)
>>> tom.color("green")
>>> tom.circle(50)
>>> tom.color("blue")
>>> tom.circle(25)
```

The figure you generated should look like Figure 3.4.

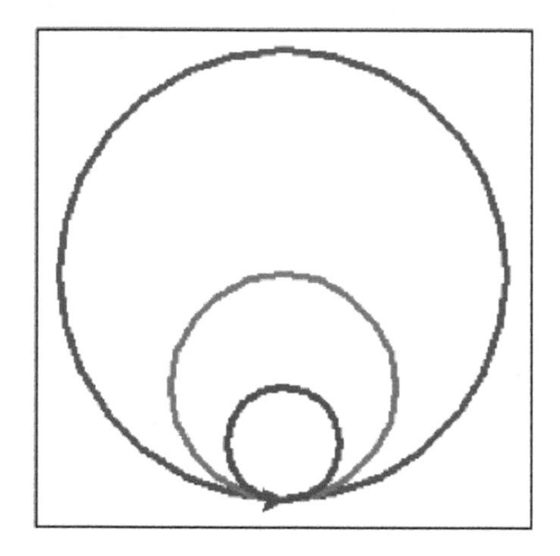

FIGURE 3.4 Coloured circles generated directly from the command line. © Stuart Smith. Used with permission.

What you have just done is:

- Get the Python module that allows you to do turtle graphics.
- Create a graphic "turtle," name it "tom," and give it a pen to draw with.
- Make the pen 3 pixels wide.
- Give tom the colour red.
- Tell tom to draw a circle with a radius of 100 pixels.
- Give tom the colour green.
- Tell tom to draw a circle with a radius of 50 pixels.
- Give tom the colour blue.
- Tell tom to draw a circle with a radius of 25 pixels.

The turtle can easily be programmed to draw other common shapes of any desired size and to place them anywhere you want in the graphics window. The necessary commands will be illustrated in the following examples.

3.3.1.3 Making and saving programs

For each of the examples below you will be given a complete program to run and then modify. To create a new program text, click on File and then select New File from the dropdown menu. The editor window will appear. You can then type in the program text[2], adding, deleting, or changing text as you would in any text editor you're familiar with. To save a program, click on File and select either Save or Save As from the dropdown menu. When you save a program in Python for Windows, it will go by default into the folder

```
C:\Users\yourUserName\AppData\Local\Programs\Python\Python36-32
```

where *yourUserName* is your user name on the system. When you want to retrieve a saved program, click on Files and then select Recent Files from the dropdown menu. You will see a list of all your saved files. Click on the one you want.

3.3.2 Geometric art with Python: Examples and experiments

3.3.2.1 n-pointed star

The first example program, `prog01`, draws a star with an arbitrary number of points. To run it, click on Run and then select Run Module in the dropdown menu. If Python pops up a window telling you to save the module, click on OK.

Listing 3.4 n-pointed star

```
1  # prog01.py
2
3  import turtle
4
5  def star(aTurtle,size,sym): # The following 5 commands define ("def") how
       to draw a star.
6      for i in range(sym):       # Do the following sym times:
7          aTurtle.right(2*360/sym)  urn right (144 degrees if sym=5).
8          aTurtle.forward(size) # Draw a line segment (100 pixels long if
               size=100).
9          aTurtle.left(360/sym) # Turn left (72 degrees if sym=5).
10         aTurtle.forward(size) # Draw a line segment (100 pixels long if
               size=100).
11
12 def main():              # The following commands are the main program.
13     tom=turtle.Pen()  # Create a turtle named "tom" and give it a "pen.!"
14     turtle.bgcolor("beige") # Set the background color to beige (Boring!
           We'll change this later)
15     tom.color("blue") # Set the pen color to blue.
16     tom.pensize(3)     # Make the lines 3 pixels wide.
17     tom.speed(3)       # Sedrawing speed slow so you can see what's
           happening.
18     size=100           # Set the length of line segments to 100 pixels.
```

[2] Python uses indentation to indicate the logical structure of a program. It is very important to indent lines *exactly* as shown in a program text (use the *Tab* key to indent).

FIGURE 3.5 Blue on beige star drawn by prog01. © Stuart Smith. Used with permission.

```
19      sym=5                 # Set the number of points on the star to 5.
20      tom.hideturtle()      # Make the turtle invisible. We don't need to see it.
21      tom.up()              # Raise the pen.
22      tom.goto(50,50)       # Move the turtle to its starting point (x=50, y=50).
23      tom.down()            # Put the pen down so the turtle can draw.
24      star(tom,size,sym)# Execute the "star" procedure defined above.
25  main()                    # End of main program.
```

You should see the turtle draw a five-pointed star (see Figure 3.5). The comments in the program (i.e., text following the "#" symbol) give a step-by-step description of what the program does. Notice that the first chunk of code (star) is a definition ("`def`") that tells *how* to make a star but does not actually cause one to be made. That happens right at the bottom of the second chunk of code ("`main`"), where you see `star(tom,size,sym)`. This command tells a turtle named "`tom`" to draw a star (here with line segments 100 pixels long and 5 points).

Experiments. In the edit window, change the values of one or more of the inputs to the program and then see what happens when you run the program again. For example, you can change the length of the line segments (`size`) or the number of points the star will have (`sym`). You can also change the background colour (`bgcolor`) or the pen colour (`color`)[3]. Try changing the turtle's drawing speed (speed). speed can range from *0* (fastest) to *10*. *6* is considered "normal," and *3* is "slow."

[3]Python knows all 140 standard web browser colours. You can find their names and sample swatches at `https://www.w3schools.com/colors/colors_names.asp`

Troubleshooting

Modifying a program brings with it the possibility of errors. Identifying and correcting them can be the most frustrating aspect of learning a new programming language. Since you usually only make small changes to the example programs here, the most likely errors will be typos. Using bright red font, Python will identify the line in the program where an error has been detected and indicate what is wrong. Typically, the last line of the error report will be the most informative. Check out that line and make sure you have not misspelled something, used lower-case where upper case is required, omitted a right parenthesis to match each left parenthesis, omitted quotation marks where they are required, etc. If you already have experience programming with another language, you know that you can also encounter errors when a running program attempts forbidden operations such as division by zero or using a variable that is undefined. The ability to fix a malfunctioning program is a skill that comes with practice, but even professional programmers sometimes stare at code for hours trying to find a problem that they know is right before their eyes. So do not feel bad if you struggle a bit when something goes wrong.

3.3.2.2 Recursive star

A single star is not a terribly exciting image, but if just two new commands are added to `prog01` a more complex image can be produced. In `prog02` the definition that tells how to make the star has been named "rstar" because it uses a powerful programming capability called "recursion." When the main program gives the command `rstar(tom,size,sym)`, the turtle makes the same sequence of left and right turns and forward moves as in `prog01`; however, each time it completes one point of the star, `rstar` commands *itself* to draw another star–but with a key difference: The length of each line segment is cut to one-third its original length. This process makes smaller and smaller stars until `size` (the length of the line segments) is 4 or fewer pixels. At this point recursion stops and no smaller stars are drawn. Run `prog02` as you did for `prog01`.

Listing 3.5 Recursive star

```
1  import turtle
2
3  def rstar(aTurtle,size,sym):
4    if size>4:   # draw a star if size is greater than 4 pixels.
5        for i in range(sym):
6            aTurtle.right(2*360/sym)
7            aTurtle.forward(size)
8            aTurtle.left(360/sym)
9            aTurtle.forward(size)
10           rstar(aTurtle,int(size/3),sym) # rstar "calls" itself here.
11
12 def main():
13   tom=turtle.Pen()
14   turtle.bgcolor("navy")
```

FIGURE 3.6 Original star elaborated recursively by prog02. © Stuart Smith. Used with permission.

```
15      tom.color("yellow")
16      tom.pensize(1)  # Reduce pen size to allow greater detail.
17      tom.speed(0)
18      size=200
19      sym=5
20      tom.hideturtle()
21      tom.up()
22      tom.goto(250,50)    # Start the turtle at (250,50) to fit on the screen
23      tom.down()
24      rstar(tom,size,sym)
25  main()
```

Figure 3.6 shows the resulting star. Notice that there are three sizes of stars and that the smaller stars are on alternating sides of the line segments of the next larger star.

Experiments. The experiments with prog02 are the same as with prog01: Change colours, size, number of points, drawing speed, pen size, etc. Some of these changes may create an image that is too large to fit in the graphics window or that disappears off one edge of the window. If this occurs you can resize the screen and then use the horizontal and vertical scrollbars to make the hidden parts of the image visible. First determine the current size of the screen and then request a larger screen (you have to estimate the required size). Type the commands following the >>> prompt. (It is not necessary to type the comments following the "#" in the commands). Python will respond by printing the width and height of the screen in pixels.

FIGURE 3.7 7-pointed star generated by modifying prog02. © Stuart Smith. Used with permission.

```
>>> turtle.screensize()          # Get the current screen size.
(400,300)
>>> turtle.screensize(2000,1500)  # Increase screen size 5X.
(2000,1500)
```

You can now move the scrollbars to find the hidden parts of your image. Figure 3.7 was made by changing the value of `sym` in `prog02` to 7. The increase in the number of points on the star increased the size of the resulting figure, which necessitated an increase in screen size.

The next part provides another example of a code written by Stuart Smith in Python, this time to create colouring book pages.

3.3.3 Geometric art with Python: Adult colouring book series

By Stuart Smith
University of Massachusetts Lowell

The "adult colouring book series" is currently a set of four Python programs that generate intricate geometric line drawings from a small number of inputs. These drawings can be kept as-is or filled with colours by hand or by using a graphics program[4]. The figures produced generally have rotational

[4]Python saves the figures as Postscript files. It may therefore be necessary to convert them to a format that can be manipulated by, e.g., MS Paint. GIMP and IrfanView (both free) are good candidates for this purpose. xrandom, xpal, and xfamily save the files in the folder where Python resides unless the user changes the destination.

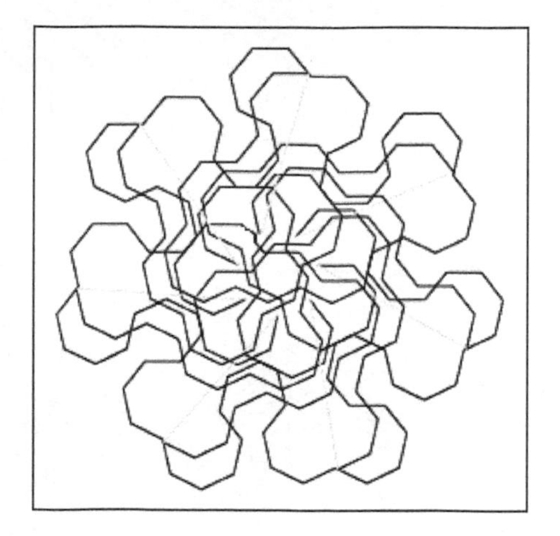

FIGURE 3.8 Python: Program output #1. © Stuart Smith. Used with permission.

symmetry, although the algorithm occasionally produces figures with translational symmetry. The basic contour of each drawing is controlled by a bit-string that can be generated either randomly or according to any of 75 cellular automaton rules. For example, the following inputs to program xrandom produce three different figures, one of which is shown in Figure 3.8.

```
1  symmetry        7
2  fraction        0.5
3  contour length  32
4  segment size    20
5  iterations      3
```

The programs are:

1. **xlines**: This program always uses a random bit string. It will generate one figure from the bit-string after it has been subjected to a specified number of applications of the cellular automaton rule.

2. **xrandom**: This program generates a set of figures from a random bit-string.

3. **xpal**: This program generates a set of figures from a palindromic bit-string.

4. **xfamily**: This program starts with the standard bit-string[5] and then generates a sequence of related figures, each representing a step in the

[5]The standard bit-string is all 0's except for a single 1 in the middle position.

evolution of the cellular automaton. In the lower left corner of each figure is a list of the parameters used to generate the figure.

The programs prompt for each input. Depending on the program chosen, a subset of the following inputs will be required:

1. **type**: 1 for the standard bit-string, 2 for a random bit string
2. symmetry: The desired order of symmetry. Symmetry should be greater than 3 and preferably a prime number. The actual symmetry obtained with a non-prime will generally be one of the prime factors of the number entered. For example, symmetry of 12 will often produce a figure with symmetry of 3.
3. **rule**: The rules are numbered 0 through 74. Pick any one and see what happens.
4. **generations**: The number of times the selected rule is applied to the bit-string
5. **contour length**: The number of line segments in the figure's contour. A figure repeats the contour n times, where n is the symmetry of the figure. Each time the contour is drawn, it is rotated 360/n degrees from its previous orientation.
6. **fraction**: When a random bit-string is used, fraction specifies the relative number of 1's in the bit-string. If fraction is set to 0.5, the resulting contour will contain approximately the same number of right and left turns. Values greater or smaller than 0.5 will bias the contour in favor of more turns in one direction than the other. $0 < \text{fraction} < 1$
7. **segment** size: The length in pixels of each line segment of the figure
8. **iterations**: The number of figures to be generated

Getting to know the range of input values that will work takes a bit of experimentation. Some general rules are:

1. symmetry can be any positive integer greater than 3, but for best results choose a prime number less than 32.
2. contour length has to be at least 2. Lengths up to 100 are generally OK; however, a large contour length together with a large symmetry and/or large segment size may produce a figure that will not fit in the drawing area.

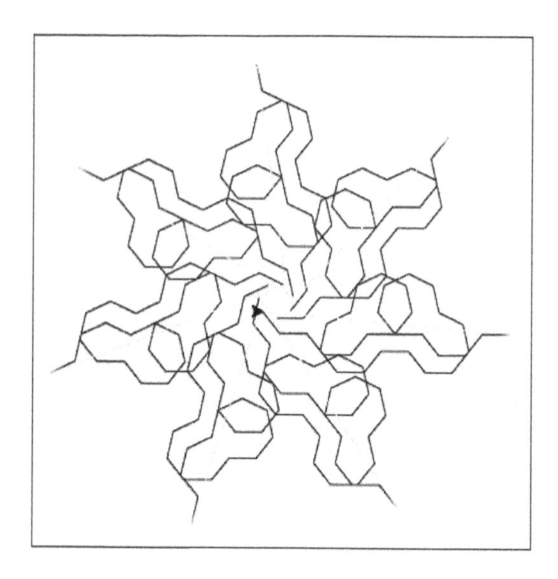

FIGURE 3.9 Python: Program output #1. © Stuart Smith. Used with permission.

3. segment size can be any integer greater than 2, but generally should be at least 10. segment size greater than 50 will often drive the "turtle" right off the drawing area if either symmetry or contour length, or both, are large.

4. generations – this value should be no more than half the value of contour length if you want to see the pure effect of each step in the evolution of the cellular automaton. For values of generations greater than half the value of contour length, the bit-string may wrap around into itself. This is perfectly OK if you like the resulting figures. See Figure 3.9 for an example.

The programs are available on the book website.

3.3.4 Algorithmic generation of design patterns with MATLAB

By Stuart Smith
University of Massachusetts Lowell

This section describes a set of MATLAB® functions that generate intricate geometric figures from a small set of numbers that determine the geometric characteristics of the figures. The kinds of figures that can be produced are reminiscent of mandalas, rose windows, wreaths, Spirograph pictures, and

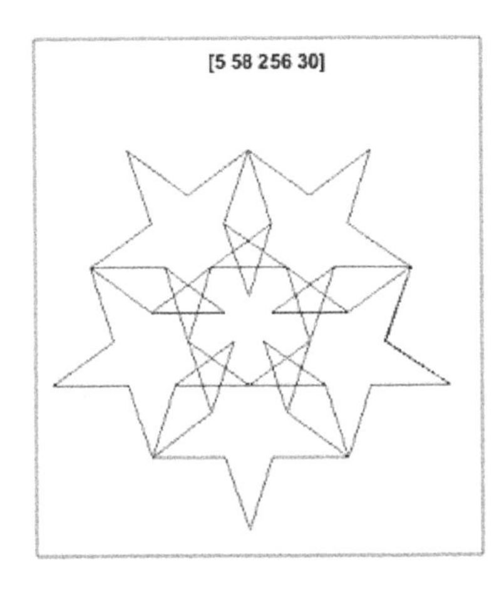

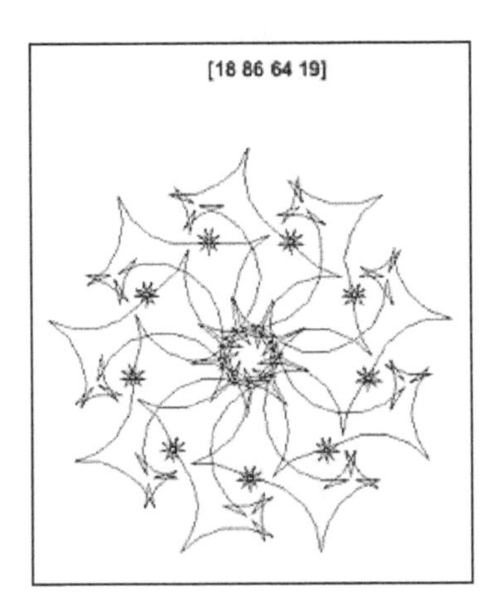

FIGURE 3.10 MATLAB: Algorithmic generation. © Stuart Smith. Used with permission.

other designs that have rotational symmetry. The outputs in Figure 3.10 are two typical examples:

The numbers at the top of each picture are the inputs to the pattern-generating routine. There are only four numbers per picture. No other inputs are needed. The four numbers are:

- **symmetry**: Symmetry is the number of repetitions of the contour (`symmetry`$\geq$ 5). Each design has a pattern of lines, called the "contour," that is exactly repeated at equally-spaced intervals around its centre.
- **rule**: The contour is developed from a list of 1's and 0's that can be modified according to any of 256 different "rules." The rules are numbered 0 through 255.
- **length**: This is the number of points (typically 16, 64, or 256) that will be joined together by line segments to form the contour.
- **steps**: This is the number of times the contour list is to be modified according to the selected rule (`steps`$\geq$ 0). Varying the number of steps is a way to generate different pictures when `symmetry`, `rule`, and `length` are held constant.

3.3.4.1 *autoxshapes*

The MATLAB function **autoxshapes** generates multiple pictures with minimal input from the user. You can begin experimenting with **autoxshapes** with no further knowledge of its inner workings. For each of the four inputs it will compute random values that are within ranges known to generally give good results. To run the program, enter for example

```
autoxshapes( 25 , 1 )
```

This will generate 25 pictures and save them on the screen in individual windows (if you do not want to save the pictures, change the second number from 1 to 0). The program will pause for 2 seconds between pictures so you have a chance to look at them. A picture saved on the screen can be written to a file: On the *picture's* menu bar click FILE, click Save As, and then name the file and save it as you would any other file on your system (if you want to manipulate a saved picture with some other graphics program, save the file in one of the common graphics file formats such as .bmp, .png., or .jpg). *NB:* A sonic representation of the contour will accompany each picture generated. If you do not want this, comment out the `Tones( B , sym )` line at the bottom of the definition of **xshapes** (i.e., put a "%" at the beginning of the line).

3.3.4.2 *xshapes*

Once you have some experience with **autoxshapes**, you can try making pictures with **xshapes**. **xshapes** generates one picture according to the values you supply for its four inputs. For example, if you enter **xshapes(11,30,64,21)** Figure 3.11 will be generated.

Figure 3.11 has the following characteristics:

- It has 11-fold symmetry (count the stars on the figure's periphery).
- Its contour is generated by rule 30.
- Its contour has 64 points.
- The contour is the result of 21 applications rule 30.

3.3.4.3 *How it works*

xshapes works in a way similar to the "turtle" graphics in Python (and earlier in the LOGO programming language). **xshapes** computes a set of points that determine the heading and distance to be travelled by the graphical turtle. The points are arrayed in a symmetrical pattern around the center of the figure to be drawn. The difference is that xshapes does not immediately

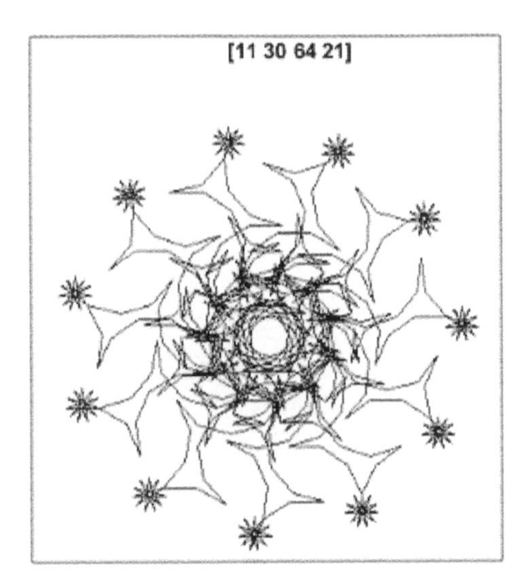

FIGURE 3.11 **xshapes** output. © Stuart Smith. Used with permission.

connect the points with line segments as it computes them, while with Python turtle graphics you can actually watch the turtle connect the points as they are computed. The reason that xshapes does not immediately connect the points is that they can be connected in a variety of interesting ways after they have been computed. The rcoord function systematically modifies the order in which the points are to be connected. rcoord temporarily skips over certain points, but it comes back to connect them later. When the points are finally connected, the result is often a surprising variety of shapes and angles. Figure 3.12 shows the array of points for Figure 3.11 before they are connected by line segments. Notice that the stars around the periphery of Figure 3.11 are constructed from small circles of points. It is easily seen that these points are not connected to their immediate neighbours in Figure 3.11 but rather to points across the circle.

3.3.4.4 Some qualifications

The description of **xshapes** above glosses over a couple of program behaviours that should be noted.

1. Sometimes **xshapes** will double the length of the contour by adding a copy of it to the end of the original contour. In this case the actual number of line segments will be twice the number specified. This is

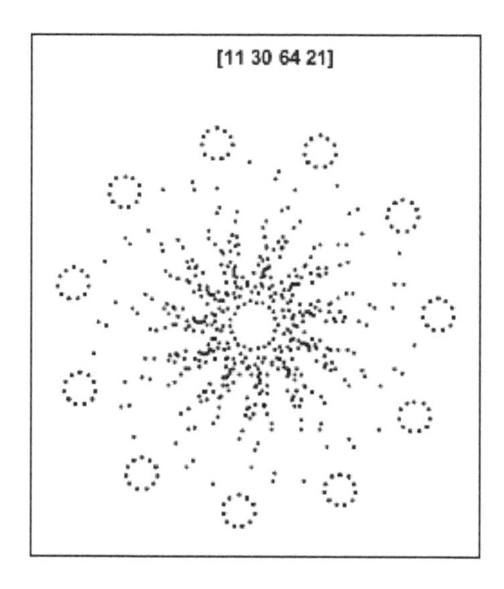

FIGURE 3.12 **xshapes** points. © Stuart Smith. Used with permission.

necessary to ensure that the contour ends up being closed (i.e., the last point is the same as the first).

2. If the symmetry specified for a figure is a non-prime number, the figure may have a lower order of symmetry. For example, the symmetry of Figure 3.10-right above was specified to be 18; however, it is obvious that the actual symmetry is 9, which is one of the factors of 18. In general, the actual symmetry of a figure whose requested symmetry is non-prime will often be one of the factors of the specified symmetry. If desired, this situation can be avoided simply by always specifying a prime number for symmetry.

3. It is possible for the computations performed by `xshapes` to simply displace the contour to the right or left, up or down, instead of rotating it around a central point. The resulting figures are rarely visually appealing.

4. The actual number of line segments in a figure's contour may be substantially smaller than the number specified. Figure 3.10-left clearly does not have 256 line segments in each repetition of the 5-pointed star pattern. This behaviour occurs when the contour computations repeatedly generate exactly the same set of points. `xshapes` will automatically remove certain redundant sets of points, but in other cases it will simply redraw the picture several times.

3.3.4.5 Under the hood

The complete design pattern application consists of more than two dozen MATLAB functions. There is no need to study this code in order to use the application successfully; however, a curious user might want to look at the overall sequence of steps that xshapes must go through to produce a picture. The internal comments identify the role each function plays.

3.3.4.6 Challenges and projects

1. Look at the MATLAB help for the plot function (enter `help plot`). Do you see how you can change the line colour or line style of xshapes pictures? Hint: Look at definition of the `Mplot` function in the SHAPES folder.

2. Look at the MATLAB help for the fill function (enter `help fill`). Do you see how you can use it to fill closed areas of an xshapes picture with colour? Hint: Look at the `Mfill` function in the SHAPES folder. Un-comment the call to `Mfill` in `xshapes` and comment out the other display functions (i.e., put a "%" sign at the beginning of a line in the program to comment it out; remove the "%" at the beginning of the line you want to be executed.) What kinds of pictures do you get now? *NB.* The MATLAB fill function cannot handle some of the more complex figures that `xshapes` can generate, so only try this exercise with fairly simple figures. Use small values for length (≤ 8).

3. After you complete #2, try the `sfill` function. It takes the same four inputs as `xshapes`. Use small values for length (≤ 8). What problems arise when length is 16 or greater?

4. If you have access to an image-processing program (e.g., GIMP), see if you can smooth out the pixelised character of `xshapes` images with a blur or anti-aliasing function.

5. Try to add background and fill colours to an `xshapes` image with, e.g.: MS Paint, Paint 3D, etc. First save the image in a file format supported by the other program: .bmp, .png, or .jpg for example.

6. In the definition of `xshapes`, comment out the call to `Mplot` and un-comment the call to `Mpoints`. When you now run `xshapes` or `autoxshapes`, what kinds of pictures do you get? Hint: See Figure 3.12.

7. If you do #5, save a hard copy of the picture. Play "connect the dots" with pencil and paper and see if you can come up with an interesting figure.

8. With a compass draw a circle. Then, with a protractor, mark off five equally spaced points on the circle (i.e., put the points $72^{\underline{o}}$ apart). Now,

going either clockwise or counter-clockwise around the circle, connect each point with a straight line to the point *after* its neighbour until you get back to the starting point. What shape do you get? The result should suggest how the function `rcoord` makes interesting shapes from points that lie on a circle.

9. After you have run `autoxshapes` or `xshapes` many times, you will probably notice that certain shapes appear very often. Perhaps the most obvious one is the star with different numbers of points. What other shapes appear frequently? Are there combinations of two or more different shapes that occur together frequently? If yes, which ones?

10. In the definition of `autoxshapes`, find the lists of numbers called `bestrules`. Pick a number between 0 and 255 that is *not* in the lists and use it for the *second* input to `xshapes` (i.e., the rule). What kind of picture does `xshapes` generate?

11. Here are two problems for mathematicians and fans of M. C. Escher's work:

    - Can you predict from the four input numbers whether or not `xshapes` will generate a shape that does *not* have rotational symmetry? If so, how?
    - When the symmetry you specify is *not* a prime number, can you predict whether an `xshapes` figure will have the symmetry you want or, instead, some factor of it? If so, how?

12. Try the different sound options at the bottom of the `Tones` function. *NB:* Un-comment only one option at a time.

The images in Figure 3.13 provide more outputs.

3.3.4.7 Getting the software and using MATLAB

This project assumes that you already have at least some experience using MATLAB. To use `xshapes` and `autoxshapes`, you need only download the SHAPES folder from the publisher's website.

Put the SHAPES folder in any convenient place on your system. It would also be a good idea to put this folder on the MATLAB "path." For this do the following:

1. Click on Set Path in the menu bar at the top of the MATLAB command window. This will take you to a list of folders already on the path.

2. Click on Add Folder. This will bring up a list of the folders on your system.

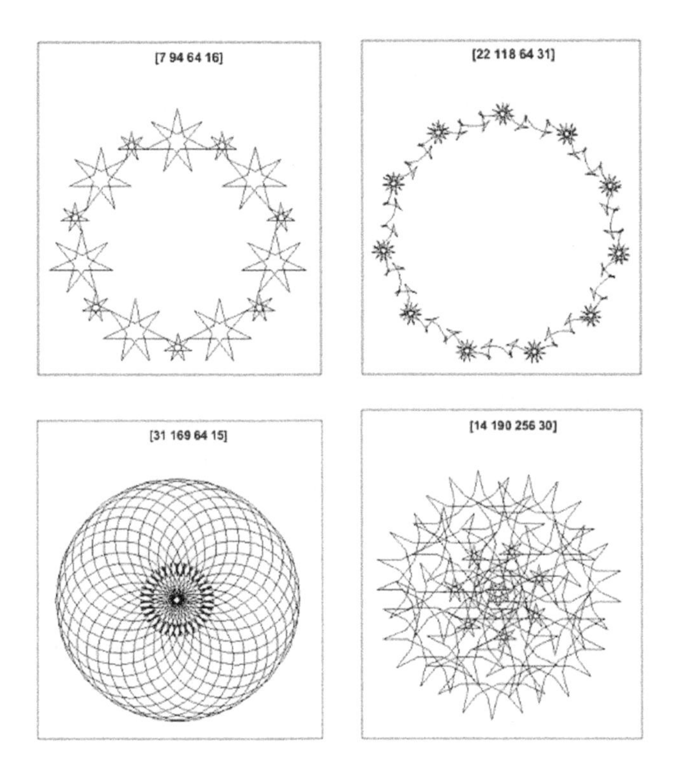

FIGURE 3.13 MATLAB: More examples. © Stuart Smith. Used with permission.

3. Find SHAPES in the list and click on it,
4. Click on Select Folder.

SHAPES will be added at the top of the list. You can use the Move Down or Move to Bottom button to put the SHAPES folder further down the list of items on the path (recommended). Save the path and Close. If you have completed all the steps above, just launch MATLAB and you will be ready to use **autoxshapes** and **xshapes** to make pictures.

3.3.4.8 Background

Gérard Langlet published the original SHAPES algorithm in 1993 [29]. It was written – in a somewhat obscure form – in APL. It has been rewritten in Dyalog APL, and at the same time a modified version that generates the contour bit-string with Wolfram's simple cellular automaton [54]. This latter version, called `xshapes`. is the basis of the MATLAB program of the same name described here. The motivation for writing `xshapes` was to replace certain *ad hoc* features of SHAPES with a consistent method based almost entirely on Boolean operations. The ability to render the contour bit-string in sound is a completely new addition to the capabilities of SHAPES-based programs. MATLAB has built-in functions to create, play, and save sounds. This made it relatively easy to implement a sound-rendering feature in the MATLAB version of `xshape`.

3.4 3D VISUALISER WITH JAVA

FROM 2D to 3D. Now that you have completed some 2D shapes let's move on to some 3D forms. The following exercise is about building your own 3D shape visualiser.

There was a time in the 1990s when any aspiring games programmer dreamt of building their own 3D render engine. This was a time when there were few choices in modelling packages (3DStudio, Blender, Alias, Wavefront and so on) or game engines (Quake, Wolfenstein, Atari, OGRE among others) to work with. It was a challenge just to fathom what was needed to draw 3D objects on monitor. Compiling languages (Java, Pascal, C) and computers powerful enough to support this were mostly proprietary (Silicon Graphics, SUN, Cray). But that did not stop them from trying. Not much has changed since then – except perhaps that there is a much greater choice of modelling and rendering software available (a lot of which is freeware). Nonetheless, in order to understand how many of today's 3D modelling, rendering and game engines work it is instructive to build your own. That is what this next project proposes to do – build your own 3D render engine using Java SE. It steps through how 3D objects are created, coloured, rendered and animated.

While there are other languages that are better at producing 3D graphics, they require a deeper understanding and lots of external libraries (OpenGL, DirectX/3D, Metal). That is why we are using Java here. Using only the standard Java class libraries, we can build a robust, but comprehensible 3D render engine in less than 20kb. In short, the simplest 3D render engine should include: 3D projection (orthographic), using triangular polygon faces to construct more complex 3D objects, colouring of those triangular polygon faces, obscuration sorting (what is in front of what), and finally rendering of the coloured triangular polygon faces so they can be seen on the computer monitor[6].

By using this simple approach to creating a 3D object and rendering it in Java, it opens up the possibility to create many different sorts of 3D objects

[6]This project was inspired by Platon Pronko's blog at: Rogach on Scala (`http://blog.rogach.org/2015/08/how-to-create-your-own-simple-3d-render.html` and `https://gist.github.com/Rogach`).

and apply effects to them. For example, if we know how to get the data from an audio feed (shown in the next project) we can use this to drive an animation for a 3D object. But first the basics – we need to learn how to construct, render and draw a 3D object in Java. Later we can feed its parameters some data from an audio visualiser. For those who want to push it a bit further, at the end of this project we have included an interesting morphing action that 'inflates' a tetrahedron into a sphere. This gives us a hint at what we could animate with our audio data in the next project.

3.4.1 Simple wireframe tetrahedron

The simplest 3D volumetric object constructable is a tetrahedron. Therefore, we are going to build a tetrahedron, first from lines, then fill in the spaces between them.

What this build does is to specify some coordinate points in 3D space and draw lines between them. Collectively they represent the sides of a tetrahedron. We can see them on our computer monitor because we have constructed a panel to paint our lines white against a black background. This build includes some sliders we can use to rotate our object in two directions.

To get a handle on this project, take a look at the code. You will notice it is inside the `paintComponent()` method that we construct our triangulated faces from vertices that make up our tetrahedron. Each vertex specifies a coordinate point in 3D space. Three vertices define a Triangle. When all four triangles are Constructed, a tetrahedron is formed. But, before we can even build our Vertices, we need a class to describe how they are constructed. We also need a class to describe how our triangles are formed from our vertices. And, finally, we need a `Matrix` class to describe how we can manipulate our triangles (so we can rotate them, etc.). We place any extra classes at the bottom of our code document. That is because when the main document is compiled these will be compiled into their own freestanding class files outside the main class file (you will see this when you compile – there will be extra class files produced).

The `Vertex` class declares three variables `x`, `y` and `z` of type `double`. It then specifies that these variables can have values added to them. The `Triangle` class declares three variables as vertices from the `Vertex` class `v1`, `v2`, and `v3` and another variable for colour. It then specifies that these four variables can have values added to them. And finally, the `Matrix` class is a three-by-three matrix class. In this class we take each coordinate point for a vertex and treat it as a vector from the origin which can be manipulated in any of three planes (x,y; x,z; y,z;). This means we either rotate or scale it (change its distance form the origin) about any of the three planes. We will use the matrix to perform rotation only at this stage. To move points around (translation) we would need a four-by-four matrix – the fourth value is the vector rotation relative to the origin.

Getting back to the main part of the code inside the **paintComponent()** method after the vertices have been constructed, we see the matrix is called to do the heading and pitch rotations. Some simple trigonometry is used to multiply the various vertex vector angles relative to the origin so that it appears to rotate all the whole triangles and thus the tetrahedron.

Next we build and add some colour to our lines and send the result to the screen buffer. Finally we call up our **renderPanel** and use **listeners** to send any changes to our vertices from the **sliders** and make our frame visible.

Listing 3.6 Tetra wireframe

```
1  import javax.swing.*;
2  import java.awt.*;
3  import java.util.List;
4  import java.util.ArrayList;
5  import java.awt.geom.*;
6
7  public class TetraWireframe {
8
9      public static void main(String[] args) {
10         JFrame frame = new JFrame();
11         Container pane = frame.getContentPane();
12         pane.setLayout(new BorderLayout());
13
14         // slider to control horizontal rotation
15         JSlider headingSlider = new JSlider(-180, 180, 0);
16         pane.add(headingSlider, BorderLayout.SOUTH);
17
18         // slider to control vertical rotation
19         JSlider pitchSlider = new JSlider(SwingConstants.VERTICAL, -90,
               90, 0);
20         pane.add(pitchSlider, BorderLayout.EAST);
21
22         // panel to display render results
23         JPanel renderPanel = new JPanel() {
24                 public void paintComponent(Graphics g) {
25                     Graphics2D g2 = (Graphics2D) g;
26                     g2.setColor(Color.BLACK);
27                     g2.fillRect(0, 0, getWidth(), getHeight());
28
29                     List<Triangle> tris = new ArrayList<>();
30                     tris.add(new Triangle(new Vertex(100, 100, 100),
31                                           new Vertex(-100, -100, 100),
32                                           new Vertex(-100, 100, -100),
33                                           Color.WHITE));
34                     tris.add(new Triangle(new Vertex(100, 100, 100),
35                                           new Vertex(-100, -100, 100),
36                                           new Vertex(100, -100, -100),
37                                           Color.WHITE));
38                     tris.add(new Triangle(new Vertex(-100, 100, -100),
39                                           new Vertex(100, -100, -100),
40                                           new Vertex(100, 100, 100),
41                                           Color.WHITE));
42                     tris.add(new Triangle(new Vertex(-100, 100, -100),
43                                           new Vertex(100, -100, -100),
44                                           new Vertex(-100, -100, 100),
45                                           Color.WHITE));
```

```
46
47                  double heading =
                        Math.toRadians(headingSlider.getValue());
48                  Matrix3 headingTransform = new Matrix3(new double[] {
49                          Math.cos(heading), 0, -Math.sin(heading),
50                          0, 1, 0,
51                          Math.sin(heading), 0, Math.cos(heading)
52                      });
53                  double pitch = Math.toRadians(pitchSlider.getValue());
54                  Matrix3 pitchTransform = new Matrix3(new double[] {
55                          1, 0, 0,
56                          0, Math.cos(pitch), Math.sin(pitch),
57                          0, -Math.sin(pitch), Math.cos(pitch)
58                      });
59                  Matrix3 transform =
                        headingTransform.multiply(pitchTransform);
60          // heading and pitch trasnforms multiplied so that they work
                together
61
62          g2.translate(getWidth() / 2, getHeight() / 2);
63
64          g2.setColor(Color.WHITE);
65          for (Triangle t : tris) {
66            Vertex v1 = transform.transform(t.v1);
67            Vertex v2 = transform.transform(t.v2);
68            Vertex v3 = transform.transform(t.v3);
69            Path2D path = new Path2D.Double();
70            path.moveTo(v1.x, v1.y);
71            path.lineTo(v2.x, v2.y);
72            path.lineTo(v3.x, v3.y);
73            path.closePath();
74            g2.draw(path);
75          }
76
77                }
78            };
79          pane.add(renderPanel, BorderLayout.CENTER);
80
81          headingSlider.addChangeListener(e -> renderPanel.repaint());
82          pitchSlider.addChangeListener(e -> renderPanel.repaint());
83
84          frame.setSize(400, 400);
85          frame.setVisible(true);
86      }
87
88  }
89
90  class Vertex {
91      double x;
92      double y;
93      double z;
94      Vertex(double x, double y, double z) {
95          this.x = x;
96          this.y = y;
97          this.z = z;
98      }
99  }
100
101 class Triangle {
102     Vertex v1;
103     Vertex v2;
104     Vertex v3;
105     Color color;
106     Triangle(Vertex v1, Vertex v2, Vertex v3, Color color) {
107         this.v1 = v1;
108         this.v2 = v2;
109         this.v3 = v3;
```

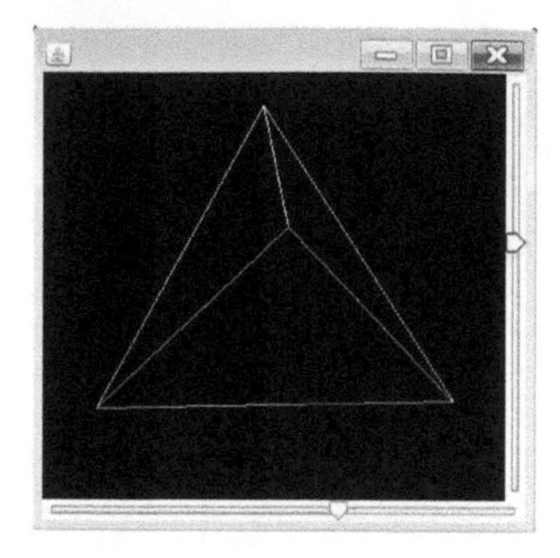

FIGURE 3.14 Tetra wireframe. © Theodor Wyeld. Used with permission.

```
110            this.color = color;
111        }
112    }
113
114    class Matrix3 {
115        double[] values;
116        Matrix3(double[] values) {
117            this.values = values;
118        }
119        Matrix3 multiply(Matrix3 other) {
120            double[] result = new double[9];
121            for (int row = 0; row < 3; row++) {
122                for (int col = 0; col < 3; col++) {
123                    for (int i = 0; i < 3; i++) {
124                        result[row * 3 + col] +=
125                            this.values[row * 3 + i] * other.values[i * 3 +
                                 col];
126                    }
127                }
128            }
129            return new Matrix3(result);
130        }
131        Vertex transform(Vertex in) {
132            return new Vertex(
133                in.x * values[0] + in.y * values[3] + in.z * values[6],
134                in.x * values[1] + in.y * values[4] + in.z * values[7],
135                in.x * values[2] + in.y * values[5] + in.z * values[8]
136            );
137        }
138    }
```

After compiling and running this code you should see something like Figure 3.14.

Note: Use control + c to close a Java session if it is running from the command prompt window.

3.4.2 Filled coloured tetrahedron

Now that we have our tetrahedron shape constructed and working as a wireframe it is only few steps away from being fully filled and coloured. In order to fill our triangles and specify a colour for the fill, first we need to use an image buffer (somewhere to store our images as our object moves and is updated by our program). When only the lines between vertices was being drawn to the screen, the renderer was able to do this by simply redrawing the lines at whatever frame-rate the computer is running. However, filling the triangles with colour and constantly redrawing would overtax the processor. Hence, it is necessary to buffer the image and only redraw if there is a change in the scene (such as rotation of our object).

To buffer the image, make sure you have imported the standard Java class for buffering images at the top of your code:

```
import java.awt.image.BufferedImage;
```

Next we need to create a buffered image inside our `paintComponent()` method. To do this add the `BufferedImage img` definition just after the two `Matrix3` transforms have been multiplied:

```
Matrix3 transform = headingTransform.multiply(pitchTransform);

BufferedImage img = new BufferedImage(getWidth(), getHeight(),
BufferedImage.TYPE\_INT\_ARGB);
```

Now we need to do the calculations: Find where the pixels are on each triangle, compute a rectangular pixel array for the bounds of the coloured area (pixels are square and even in a triangular shape they are just blocks stacked on top of each other to form the overall shape). Here is the code for these calculations:

Listing 3.7 Bounds of coloured area

```
1
2  for (Triangle t : tris) {
3    Vertex v1 = transform.transform(t.v1);
4    v1.x += getWidth() / 2;
5    v1.y += getHeight() / 2;
6    Vertex v2 = transform.transform(t.v2);
7    v2.x += getWidth() / 2;
8    v2.y += getHeight() / 2;
9    Vertex v3 = transform.transform(t.v3);
10   v3.x += getWidth() / 2;
11   v3.y += getHeight() / 2;
12
13   Vertex ab = new Vertex(v2.x - v1.x, v2.y - v1.y, v2.z - v1.z);
14   Vertex ac = new Vertex(v3.x - v1.x, v3.y - v1.y, v3.z - v1.z);
15   Vertex norm = new Vertex(ab.y * ac.z - ab.z * ac.y, ab.z * ac.x
16        - ab.x * ac.z, ab.x * ac.y - ab.y * ac.x);
17   double normalLength = Math.sqrt(norm.x * norm.x + norm.y * norm.y
18        + norm.z * norm.z);
19   norm.x /= normalLength;
```

```
20      norm.y /= normalLength;
21      norm.z /= normalLength;
22
23      double angleCos = Math.abs(norm.z);
24
25      int minX = (int) Math.max(0,
26          Math.ceil(Math.min(v1.x, Math.min(v2.x, v3.x))));
27      int maxX = (int) Math.min(img.getWidth() - 1,
28          Math.floor(Math.max(v1.x, Math.max(v2.x, v3.x))));
29      int minY = (int) Math.max(0,
30          Math.ceil(Math.min(v1.y, Math.min(v2.y, v3.y))));
31      int maxY = (int) Math.min(img.getHeight() - 1,
32          Math.floor(Math.max(v1.y, Math.max(v2.y, v3.y))));
33
34      double triangleArea = (v1.y - v3.y) * (v2.x - v3.x) + (v2.y - v3.y)
35          * (v3.x - v1.x);
36
37      for (int y = minY; y <= maxY; y++) {
38        for (int x = minX; x <= maxX; x++) {
39
40          double b1 = ((y - v3.y) * (v2.x - v3.x) + (v2.y - v3.y)
41              * (v3.x - x))
42              / triangleArea;
43
44          double b2 = ((y - v1.y) * (v3.x - v1.x) + (v3.y - v1.y)
45              * (v1.x - x))
46              / triangleArea;
47
48          double b3 = ((y - v2.y) * (v1.x - v2.x) + (v1.y - v2.y)
49              * (v2.x - x))
50              / triangleArea;
51
52          if (b1 >= 0 && b1 <= 1 && b2 >= 0 && b2 <= 1 && b3 >= 0
53              && b3 <= 1) {
54            double depth = b1 * v1.z + b2 * v2.z + b3 * v3.z;
55            int zIndex = y * img.getWidth() + x;
56            if (zBuffer[zIndex] < depth) {
57              img.setRGB(x, y, getShade(t.color, angleCos)
58                  .getRGB());
59              zBuffer[zIndex] = depth;
60            }
61          }
62        }
63      }
64
65    }
66
67    g2.drawImage(img, 0, 0, null);
```

Instead of using `g2.draw(path);` at the end of our pixel calculations we need to call up our buffered image (`img`): `g2.drawImage(img, 0, 0, null);`. Therefore, before we can compile and run, we need to remove the `g2.draw(path);` part of the code. Remove this part:

```
1
2   g2.setColor(Color.WHITE);
3   for (Triangle t : tris) {
4     Vertex v1 = transform.transform(t.v1);
5     Vertex v2 = transform.transform(t.v2);
6     Vertex v3 = transform.transform(t.v3);
7     Path2D path = new Path2D.Double();
8     path.moveTo(v1.x, v1.y);
9     path.lineTo(v2.x, v2.y);
10    path.lineTo(v3.x, v3.y);
11    path.closePath();
```

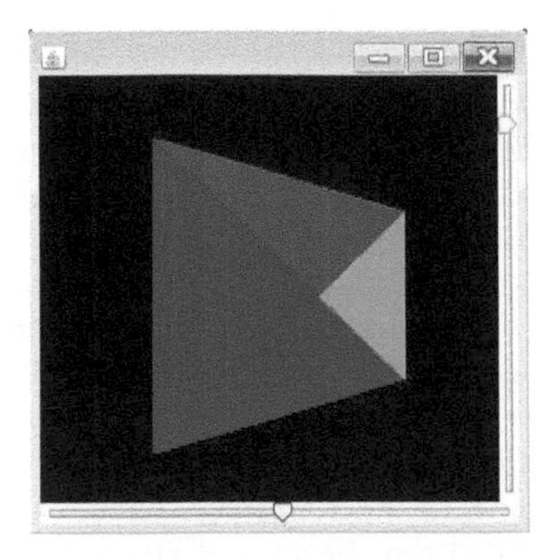

FIGURE 3.15 3D coloured bounds. © Theodor Wyeld. Used with permission.

```
12    g2.draw(path);
13  }
```

But, if we compile and run now, we notice that our tetrahedron is in the lower left corner of the panel and it is just white – not coloured. To fix its position, we need to change the `g2.translate();` values. Depending on what your `frame.setSize();` is, you need to change how much the width and height are translated. In the example code the frame is 400 × 400. As our vertices are set out at 100 pixel coordinate values we need to divide our initial frame ratios by 50. This will shift the tetrahedron shape to the centre of the panel.

Next, to ensure our triangles are actually drawn the colours we want, we need to specify a colour inside the individual vertex definitions. Inside each `tris.add(new Triangle(new Vertex( )));` there is a call to the `Color` component of the vertex class. A range of colours is available in the super `java.awt.Color;` class. Because we are importing `java.awt.*;` we have access to the logical colours (for the full list of logical colours available, see `https://docs.oracle.com/javase/7/docs/api/java/awt/Color.html`). Therefore, simply add colours to each vertex definition (remember to use capitals). For example,

```
1
2  tris.add(new Triangle(new Vertex(100, 100, 100),
3       new Vertex(-100, -100, 100),
4       new Vertex(100, -100, -100),
5       Color.RED));
6  // specifies this triangle should be draw in the colour RED
```

Now when we compile and run we should see something like Figure 3.15. The full code is available on the book's website.

However, there is clearly still a problem with our renderer. When we rotate the shape, we notice that the blue triangle appears to be in front, regardless of the orientation of the overall object. This is because the renderer is simply following the order that the triangles are specified. Each time the scene is manipulated (the object is rotated) it is redrawn according to the order specified in the code layout. To fix this, we need to do a check on what pixels are actually in front of others. In our 3D world coordinate space the y axis is up, x axis to the right and the z axis is pointing out of the screen, towards the viewer. Therefore, all we need to do is check for which parts of a triangle are farthest from the origin (or closest to the screen), and always draw them last (so they appear to be in front of whatever is behind them). This operation is called a z-buffer. It is called that because it stores the z values for each pixel in the screen and sorts them just before the final render so they get drawn in the correct order.

3.4.3 Z-Buffered coloured tetrahedron

In order to implement the z-buffer we need to create an array (buffer) to store the z values for our pixels (their ordering along the z axis of our screen space – the axis pointing out of the screen) so we can retrieve them at render time. To achieve all of this we need to add some more operations into our base code that lets us calculate all the depths of the pixels and arrange them in the buffer so they get drawn in the correct order. First we need to add the z-buffer or array generator. Place the below code just under where the `BufferedImage img` is created:

```
1  double[] zBuffer = new double[img.getWidth() * img.getHeight()];
2  //initialize array with extremely far away depths
3  for (int q = 0; q < zBuffer.length; q++) {
4    zBuffer[q] = Double.NEGATIVE_INFINITY;
5  }
```

And then we need to tell the z-buffer to do the sorting and send to the renderer:

```
1  double depth = b1 * v1.z + b2 * v2.z + b3 * v3.z;
2  int zIndex = y * img.getWidth() + x;
3  if (zBuffer[zIndex] < depth) {
4    img.setRGB(x, y, t.color.getRGB());
5    // moved here from the previous for loop
6    zBuffer[zIndex] = depth;
7  }
```

Notice this line of code has been moved inside the if (zBuffer...) { } statement:

```
img.setRGB(x, y, t.color.getRGB());
```

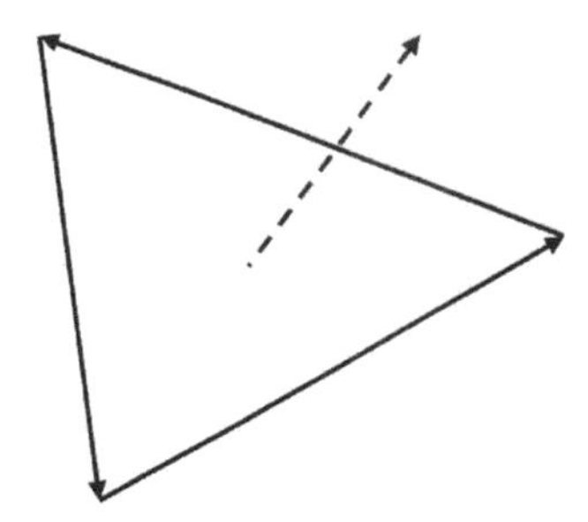

FIGURE 3.16 Concept: Normal to the plane vector. © Theodor Wyeld. Used with permission.

Now, when we compile and run we should see that the triangles are drawn in the correct order – the one that appears to be closest to the screen is drawn last, thus in front of the others. The full code is available on the book's website.

3.4.4 Cosine shaded tetrahedron

There is one more type of rendering we can apply to our tetrahedron to make it a bit more realistic (in fact there are many more (specular, anisotropic, material mapped and so on) – a cosine shader is the easiest to implement that provides for a good result. In order to simulate the way light bounces off surfaces we can simply calculate the angle of incidence between the light source and the surface. According to this angle we can lighten or darken the apparent surface colour to simulate how much of the light is being reflected. This creates what is called a colour ramp – the gentle increasing of colour brightness value across a surface. The net effect is the apparent perception of depth due to shading of an object's surface.

In order to implement the cosine shader we first need to know what direction our triangles are facing. This is called a normal to the plane vector. Basically, if you think of a vector that connects the vertices of your triangle moving from one to the next in a counter-clockwise manner, the normal vector (dashed line) is pointing away from that plane. See Figure 3.16.

Once we know what the normal direction is for each triangle we can then calculate the angle between the normal vector for each plane and the light source. In this example, we will take our light source to be behind the camera with parallel rays. Add the following code inside the `for (Triangle t : tris) { }` for loop, just under the third vertex transform:

```
1  Vertex ab = new Vertex(v2.x - v1.x, v2.y - v1.y, v2.z - v1.z);
2  Vertex ac = new Vertex(v3.x - v1.x, v3.y - v1.y, v3.z - v1.z);
3  Vertex norm = new Vertex(
4    ab.y * ac.z - ab.z * ac.y,
5    ab.z * ac.x - ab.x * ac.z,
6    ab.x * ac.y - ab.y * ac.x
7  );
8  double normalLength = Math.sqrt(norm.x * norm.x + norm.y * norm.y + norm.z
       * norm.z);
9  norm.x /= normalLength;
10 norm.y /= normalLength;
11 norm.z /= normalLength;
12
13 double angleCos = Math.abs(norm.z);
```

And, finally we need to specify what colours to add and which triangles to colour. To do this we need to insert a colouring method outside our `paintComponent()` method and after the panel has been constructed, but still inside our main class file `TetraFilledTriangles` {}:

```
1  public static Color getShade(Color color, double shade) {
2    int red = (int)(color.getRed() * shade);
3    int green = (int)(color.getGreen() * shade);
4    int blue = (int)(color.getBlue() * shade);
5
6    return new Color(red, green, blue);
7  }
```

And we need to add the cosine shader to our `img` constructor inside the `zBuffer` if statement. To do this, change:

```
img.setRGB(x, y, t.color.getRGB());
```

to:

```
img.setRGB(x, y, getShade(t.color, angleCos).getRGB());
```

Now when we compile and render we noticed that as we rotate the tetrahedron, the sides change colour value according to the angle they present to the screen or camera. But, while our eyes can detect a wide range of colours at different luminance levels, a computer monitor can never approach the same range or brightness as the sun. Therefore a Gamma curve is applied to the brightness value of colours so that they 'look' brighter. It is a bit like the loudness filter for music – a curve that boosts the bass and treble frequencies so that the music 'sounds' louder. Java's RGB colour gamut already includes a Gamma of 2.2. This means our simple Cosine shader has an attenuated or exponential dropoff when the faces are turned away from the camera. In other words, as you rotate the object it gets darker very quickly. To correct this we need to make the dropoff linear and then apply our own Gamma curve. A simple approximation of this process involves partially reversing the Gamma curve. To do this, change the `Color getShade();` method to:

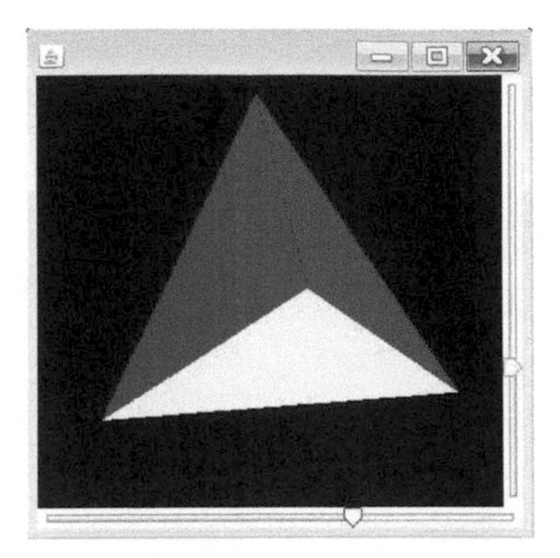

FIGURE 3.17 Faces of the planes. © Theodor Wyeld. Used with permission.

```
1  public static Color getShade(Color color, double shade) {
2    double redLinear = Math.pow(color.getRed(), 2.4) * shade;
3    double greenLinear = Math.pow(color.getGreen(), 2.4) * shade;
4    double blueLinear = Math.pow(color.getBlue(), 2.4) * shade;
5
6    int red = (int) Math.pow(redLinear, 1/2.4);
7    int green = (int) Math.pow(greenLinear, 1/2.4);
8    int blue = (int) Math.pow(blueLinear, 1/2.4);
9
10   return new Color(red, green, blue);
11 }
```

Now when we compile and render this we should see that our faces do not get dark too quickly. See Figure 3.17. The full code is available on the book's website.

3.4.5 From tetrahedron to sphere

An interesting aside to Platon Pronko's (Rogach's) blog on Scala is his inclusion of a script to 'inflate' the tetrahedron into a sphere. In the very early days of 3D animation, a technique for morphing objects from one form to another was developed as a low-cost (CPU-wise) method. Basically, as long as the two objects morphed between included exactly the same number of vertices, any shape could be morphed into any other shape (within some limitations), for example, from a rabbit to a duck (see Figure 3.18).

What Pronko has done is slightly different. He has begun by subdividing each of the tetrahedron's triangular sides into four smaller triangles, four times, to create a total of 256 triangles for each main triangle. Then the location of the vertices for these new, subdivided, triangles are given a new coordinate distance relative to the origin such that they appear to form a curve on the surface. When all four main faces of the tetrahedron are processed in this manner, a sphere is produced.

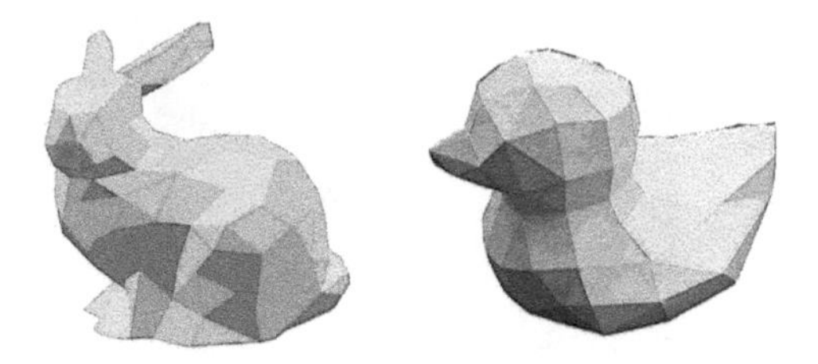

FIGURE 3.18 Rabbit to a duck. © Theodor Wyeld. Used with permission.

To add the tetrahedron-to-sphere inflation method, we start by adding in how many times the inflation will proceed (4). Place the following code between the end of the **Triangle ArrayList<>();** and the first Matrix transform (heading):

```
1  for (int i = 0; i < 4; i++) {
2    tris = inflate(tris);
3  }
```

Then, insert the below just after the **Color getShade();** method and before the final curly braces inside the main class.

```
1  public static List<Triangle> inflate(List<Triangle> tris) {
2    List<Triangle> result = new ArrayList<>();
3    for (Triangle t : tris) {
4      Vertex m1 = new Vertex((t.v1.x + t.v2.x)/2, (t.v1.y + t.v2.y)/2,
           (t.v1.z + t.v2.z)/2);
5      Vertex m2 = new Vertex((t.v2.x + t.v3.x)/2, (t.v2.y + t.v3.y)/2,
           (t.v2.z + t.v3.z)/2);
6      Vertex m3 = new Vertex((t.v1.x + t.v3.x)/2, (t.v1.y + t.v3.y)/2,
           (t.v1.z + t.v3.z)/2);
7      result.add(new Triangle(t.v1, m1, m3, t.color));
8      result.add(new Triangle(t.v2, m1, m2, t.color));
9      result.add(new Triangle(t.v3, m2, m3, t.color));
10     result.add(new Triangle(m1, m2, m3, t.color));
11   }
12   for (Triangle t : result) {
13     for (Vertex v : new Vertex[] { t.v1, t.v2, t.v3 }) {
14       double l = Math.sqrt(v.x * v.x + v.y * v.y + v.z * v.z) /
             Math.sqrt(30000);
15       v.x /= l;
16       v.y /= l;
17       v.z /= l;
18     }
19   }
20   return result;
21 }
```

When you add this code and it is compiled and run you should see something like Figure 3.19. The full code is available on the book's website.

The reason the inflate method was included in this build was because we can use it to demonstrate how morphing can be animated in real time. Once

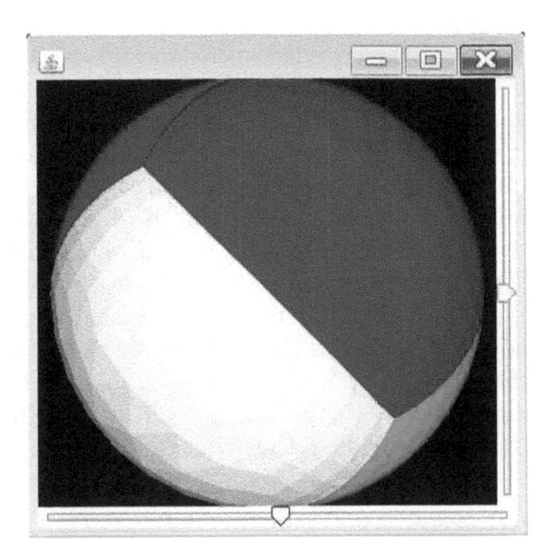

FIGURE 3.19 From tetrahedron to sphere. © Theodor Wyeld. Used with permission.

we know the parameters for our inflate method we can access these using a slider to control how our tetrahedron is inflated over time. In other words, we can animate its transformation from tetrahedron to sphere over time. But, before we build this, just to demonstrate how powerful this program is, in the next build we will change the type of object originally modelled.

3.4.6 Changing the type of initial 3D object constructed to a box

Just to recap what we have done so far, our base code includes a GUI wrapper to display the results of the render engine to the screen with sliders for rotating the 3D object about two axes, two classes, one to provide coordinates for the vertices of the 3D object called `Vertex` and another called `Triangle` which refers to the faces produced by groups of three vertices (the minimum vertex arrangement for a polygon face). From here the vertices are specified in an array which essentially provides the 3D coordinates for the shape of the final object. Once the vertices are specified in the array, lines are drawn between them using the `awt.geom.Path2D` class method. In order to rotate the 'object' it needs to be 'transformed'. To do this, a three-by-three matrix class was created to multiply values across the vertices so that they appear to change their position on the screen. The sliders feed values to the transformation matrix to effect the rotation. And, because a slider is used, a listener is also needed to cast or send the values to the `Matrix3` class. To fill the triangles with colour, only those pixels included inside the triangles are drawn to the screen. To do this, Barycentric coordinates are used, as they easily convert to Cartesian coordinates (x, y). Once the parameters of the face of a triangle were found, colour was added. But, because the draw method simply redraws whatever is presented, there needed to be a way of checking to see if one face of the 3D object is in front of the other. To do this, a z-buffer was used. It simply checks the effective z distance for each pixel in an array from the camera to see if it is greater or less than the next one. This means that it draws the pixels closest to the camera last. In order to add some shading, a simple Cosine algorithm was used. This calculates the angle between the way a light source bounces off a surface towards the viewer and builds a colour ramp on the surface at render time, making it look a bit more realistic. Finally, as a quirky addition, a method for subdividing each triangle into four smaller ones was included. The net effect is to morph a sphere from the original tetrahedron shape.

We could do a lot more with this simple renderer. But for now, it serves its purpose. However, just to add some inspiration, the following outlines how to change the initial 3D object type (tetrahedron), include some more useful sliders (heading, pitch, roll and field-of-view (FoV)), and use a four-by-four matrix to create a fully perspective view. To do all of this, we need to specify more triangular faces (to build a cube) and add an extra field to the matrix transformation class.

To specify the additional vertices needed to build a cube rather than a tetrahedron, you can use a pen and paper to plot out the coordinate points for a cube in 3D. For the 4×4 matrix, you need to add an extra field to each of the matrix transforms (heading, pitch, roll and FoV), and in the `Matrix4` class itself. When you compile and run this code you should see something like Figure 3.20. The full code is available on the book's website.

3.4.7 Morphing the 3D object from spiky tetrahedron to sphere

Based on Pronko's model in this build, we will add a slider to animate the transition from the tetrahedron shape to the sphere shape. In fact, we can go the other way too – a deflated, or spiky, tetrahedron. To make it clearer what some of the class variable definitions and functions are, some of the terms have been changed. There will be three separate Java files that, when compiled, create six separate class files (one each for the main class files and the subclasses `Matrix3x3`, `Triangle`, and `Vertex`).

The `CanvasTetraSphere` file contains the `main()` method. It is used to set the initial parameters for the sliders, name them and feed their output from the `ControlFrameTetraSphere` class to the `SpikyTetraSphere` class where the matrix transformation operations are performed. `CanvasTetraSphere` also

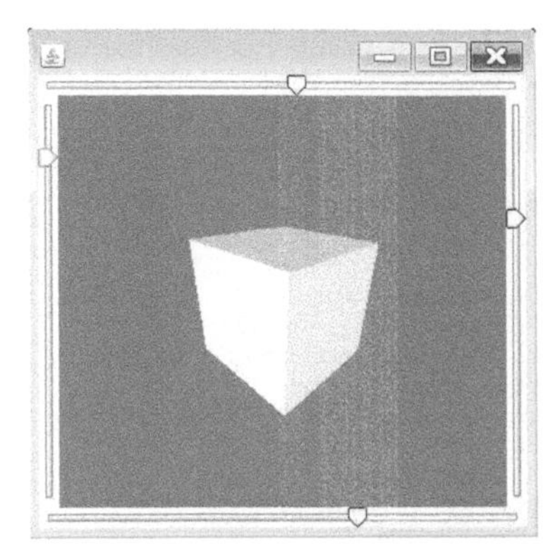

FIGURE 3.20 Changing 3D object to a box. © Theodor Wyeld. Used with permission.

sets up the `JPanel` used for displaying the 3D object and a separate panel for the controls. The `ControlFrameTetraSphere` file is basically a container for the slider controls. It directs the listener as to what to listen for. Most of the work is done in the `SpikyTetraSphere`. This is where the 3D object is constructed from the triangles array (vertex positions for each triangle used in the 3D object). It is also where the 3×3 matrix is transformed by the various values fed to it from the sliders. The z-buffering is performed here as well as the Cosine shading. And finally, the inflate and deflate methods are executed here.

3.4.7.1 Control frame

First we will put the slider controls in their own GUI class file called `ControlFrameTetraSphere`.

Listing 3.8 Control frame

```
1   class ControlFrameTetraSphere extends JFrame{
2     private JPanel container;
3     private Font headingFont = new Font("Sans-Serif", Font.BOLD, 16);
4
5     public ControlFrameTetraSphere(){
6       setTitle("Controls");
7       setDefaultCloseOperation(JFrame.EXIT_ON_CLOSE);
8       container = new JPanel();
9       container.setLayout(new BoxLayout(container, BoxLayout.Y_AXIS));
10      this.add(new JScrollPane(container));
11    }
12
13    public void addHeading(String text){
14      JLabel heading = new JLabel(text);
15      heading.setFont(headingFont);
16      heading.setAlignmentX(LEFT_ALIGNMENT);
17      container.add(heading);
18      container.add(Box.createVerticalStrut(15));
```

```
19      pack();
20    }
21
22    public void addSlideControl(String label, int min, int max, int value,
          ChangeListener cl){
23      addSlideControl(label, min, max, value, 100, cl);
24    }
25
26    public void addSlideControl(String text, int min, int max, int value,
          int tickSpacing, ChangeListener a){
27      JSlider slider = new JSlider(min, max, value);
28      slider.addChangeListener(a);
29      slider.setBackground(new Color(225, 225, 250));
30      slider.setAlignmentX(LEFT_ALIGNMENT);
31      JLabel label = new JLabel(text);
32      label.setAlignmentX(LEFT_ALIGNMENT);
33      container.add(label);
34      container.add(slider);
35      container.add(Box.createVerticalStrut(15));
36      pack();
37    }
38  }
```

3.4.7.2 3D Object and draw method class

Next we need a deflate method for reversing the inflate method. We have added a '`spherise`' and scale slider input functions and the math necessary to do the animating between tetrahedron and sphere. You will notice in the `SpikyTetraSphere` class file, we declare our slider initial values at the top of the code, build our triangular faces and add them to an `ArrayList`, check for which way the faces of our triangles are up (normal vector) and define a '`rate`' for the morphing of the object. This is used to send values from the `spherise` slider to the inflate and deflate methods.

In short, for the inflate method, if we want to find the arc length between two points on a surface we can use the formula:

$$y = \sqrt{1 - x^2} \tag{3.1}$$

Essentially, the inflate method multiplies each vertex by the arc distance between the edges of the main triangular faces of the tetrahedron and its centre as the origin. Because it increments using '`rate`' we see a gentle expansion of each tessellated face of the tetrahedron moving outwards to form a sphere. By contrast the deflate method uses a hyperbolic function to reduce the arc between points in a slightly more radical manner. See Figure 3.21. The full code is available on the book's website.

3.4.7.3 Canvas class

Finally, we need a main method to launch our program. The `CanvasTetraSphere` class file calls up the GUI for displaying the 3D object and the GUI for

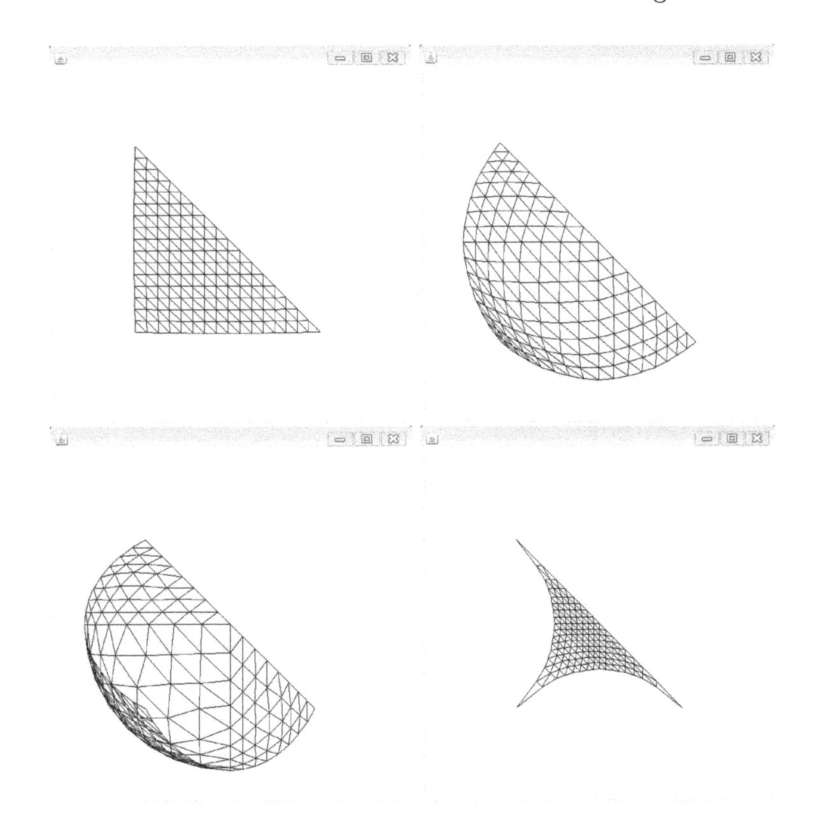

FIGURE 3.21 Various states of the inflation method: None, Partial, Fully Inflated, and Fully Deflated. © Theodor Wyeld. Used with permission.

the controls. It launches each of the class files and specifies the initial values for the slider controls. It handles the input from the controls and passes them on to the `SpikyTetraSphere` class to update the draw method (see the book's website for full code).

Listing 3.9 Canvas class

```
1   import java.awt.Dimension;
2   import javax.swing.BoxLayout;
3   import javax.swing.JComponent;
4   import javax.swing.JFrame;
5   import javax.swing.JLayeredPane;
6   import javax.swing.JPanel;
7   import javax.swing.JSlider;
8
9   public class CanvasTetraSphere extends JPanel {
10    private JLayeredPane layeredPane;
11    public static SpikyTetraSphere tetraSphere;
12
13    public CanvasTetraSphere() {
14      setLayout(new BoxLayout(this, BoxLayout.PAGE_AXIS));
15      layeredPane = new JLayeredPane();
16          layeredPane.setPreferredSize(new Dimension(600, 600));
```

```
17
18          tetraSphere = new SpikyTetraSphere();
19      tetraSphere.setBounds(0, 0, 600, 600);
20
21      layeredPane.add(tetraSphere);
22      add(layeredPane);
23
24      ControlFrameTetraSphere controlFrame= new ControlFrameTetraSphere();
           //#
25      initControlComponents(controlFrame, tetraSphere);
26      controlFrame.setBounds(1250, 50, 300, 300);
27      controlFrame.setVisible(true); //#
28    }
29
30    public static void createAndShowGUI() {
31      JFrame frame= new JFrame();
32      frame.setDefaultCloseOperation(JFrame.EXIT_ON_CLOSE);
33      frame.setLocation(600, 50);
34      //frame.setResizable(false);
35
36      JComponent newContentPane = new CanvasTetraSphere();
37          newContentPane.setOpaque(true);
38          frame.setContentPane(newContentPane);
39
40      frame.pack();
41      frame.setVisible(true);
42    }
43
44    public static void main(String[] args) { //#
45      javax.swing.SwingUtilities.invokeLater(new Runnable() {
46              public void run() {
47                  createAndShowGUI();
48              }
49          });
50    }
51
52
53  public static void initControlComponents(ControlFrameTetraSphere
         controlFrame, SpikyTetraSphere tetraSphere) {
54      controlFrame.addHeading("TetraSphere Controls");
55
56    controlFrame.addSlideControl
57      ("Spherise", 0, 4000, tetraSphere.Spherise,  a -> { //850, 1050
58        tetraSphere.Spherise = ((JSlider)a.getSource()).getValue();
59        tetraSphere.repaint();
60      });
61
62    controlFrame.addSlideControl
63      ("Scalise", 100, 600, tetraSphere.Scale,  a -> { //100, 600
64        tetraSphere.Scale = ((JSlider)a.getSource()).getValue();
65        tetraSphere.repaint();
66      });
67
68    controlFrame.addSlideControl
69      ("Heading", -180, 180, tetraSphere.Heading,  a -> {
70        tetraSphere.Heading = ((JSlider)a.getSource()).getValue();
71        tetraSphere.repaint();
72      });
73
74    controlFrame.addSlideControl
75      ("Pitch", -180, 180, tetraSphere.Pitch,  a -> {
76        tetraSphere.Pitch = ((JSlider)a.getSource()).getValue();
77        tetraSphere.repaint();
78      });
79
80    }
81  }
```

Note: When compiling multiple Java files in a single CMD session use `javac *.java`. This is necessary if the class files cross-reference each other. If they are not compiled together, errors occur when one file cannot find another.

When all three class files have been compiled and run you should see something like that in Figure 3.22 (showing each state of the '`spherise`' slider).

3.5 AUDIO DATA TO DRIVE 3D MORPHING ANIMATION

Now that we can see how a slider feeds a data stream to the parameters for our 3D object –morphing it from a sphere to a tetrahedron to a spiky tetrahedron – we can substitute the slider data for audio data. In other words, we can use data from a music file to animate our 3D object. From chapter one, we know how to get the data from an audio file to draw a waveform on the screen. To use this same data to drive our 3D animation is similar but a bit more involved. The first thing to consider is the way the 2D animation was able to keep up with the data stream. There are some optimisations we could perform on the 2D animation, but essentially, modern computers are able to process and display the data fast enough that we do not notice too much distortion from missed frames if the data feed rate is too high. However, with a 3D object animation, there is a bigger overhead – there is more data to process and draw. To begin with, for a fully rendered 3D object we need to use the z-buffer to determine which pixels are in front of others. Then we need to calculate the angle of the face in relation to the view in order for the renderer to apply the correct Cosine shader colour ramp. All of this means we cannot rely on the machine to do the audio data processing and drawing at just any speed. Otherwise, we will see a lot of distortion, as the display cannot keep up with the data feed. There are a number of different ways to make sure our audio data-driven morphing animation is smooth: We can use a low-pass filter (such as a moving average) to smooth the peaks and troughs of the audio data thus removing some of the noise; collect and store all the audio data in an array, divide the array into smaller chunks and place the chunked samples into 'buckets', and sum and divide the values in each bucket to get an average, and then feed the new averages to our animated object at 30fps; or a combination of these (among other strategies). It is the combination that we will implement in this build: Store the sound data in an array as it becomes available but start feeding it to the 3D object's morphing parameters as soon as it is available at 30fps. By 'throttling' the data feed to 30fps we should overcome the double overhead of simultaneously processing the sound data and 3D drawing process. Basically, we are dividing the total length of our original audio file by a number that effectively sends only 30 pieces of data every second. We will talk more about that later. First we can build a basic sine wave generator and use this to drive our animation. From this we will see how a simple audio data steam can be used to drive the animation. Then

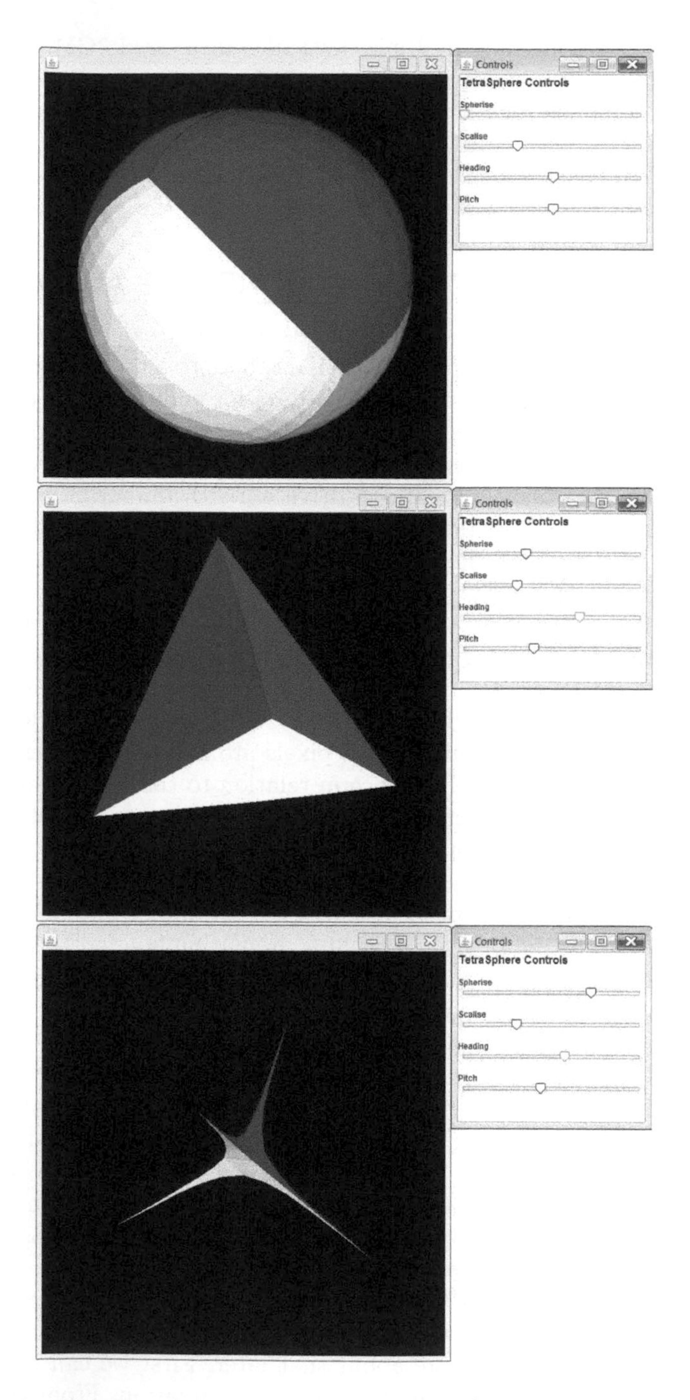

FIGURE 3.22 State of the 'spherise' slider. © Theodor Wyeld. Used with permission.

we will revisit the process of feeding an external audio file's dataline to the animation.

3.5.1 Basic sine wave generator

Before we actually feed our new method an external audio file dataline, to demonstrate how it works, we can use a sine wave generator in the first instance. The following outlines a basic sine wave generator. In short, a sine wave is a continuously changing, but consistently changing, value over time forming a periodic curve or waveform. It represents three values only: Amplitude (height) and frequency (how often it changes) over time (the intervals of change). It is represented by the formula:

$$y(t) = \text{Amplitude}.sin(2\pi.\text{frequency.time} + \text{phase}) \tag{3.2}$$

This can be read as: Values on the y axis over time (represented by the `x` axis) can be evaluated to the values amplitude multiplied by the *sin* of two × Pi × frequency × time + phase (initial position).

This can be further simplified to:

$$y(t) = \text{Amplitude}.sin(\text{Angle} + \text{Volume}) \tag{3.3}$$

For our purposes we can code this as:

$$\text{Angle} = \text{SampleRate.frequency}.\pi \tag{3.4}$$

$$\text{sine values} = sin(\text{angle}).\text{volume} \tag{3.5}$$

Using these formulas we can generate a rising and falling value.

There are simpler ways to generate a sine wave. The method used here incorporates a dataline so that we can hear the sine wave as a sound and thus is closer to our final goal of using an audio file to animate our 3D object.

In the example here we have set the sample rate to 44.1 KHz to simulate the sort of sample rate we get in a normal audio file. In fact, we are using the `javax.sound.sampled` super class to generate a dataline. We need to store the data generated by the dataline in a buffer so we can then pass it on to our draw method. You will see we sequentially open, start, drain, stop and close the dataline. This is so that it does not continue to use computing resources after its job is done. Our class file uses three for loops: One to generate our sine values, another to set up the frame to fit the drawn samples, and a final loop to generate the coordinates for dots to draw on our graph.

Listing 3.10 Draw sine wave

```
1  import javax.sound.sampled.AudioFormat;
2  import javax.sound.sampled.AudioSystem;
3  import javax.sound.sampled.SourceDataLine;
4  import javax.sound.sampled.LineUnavailableException;
5
6  import java.awt.*;
7  import java.awt.geom.*;
8  import javax.swing.*;
9
10 public class drawSinewave extends JPanel {
11
12   static double[] sinesArray;
13   static int vol;
14
15   public static void main(String[] args) {
16
17     try {
18       drawSinewave.createSinewave(200, 100);
19     } catch (LineUnavailableException lue) {
20       System.out.println(lue);
21     }
22
23     // Frame object for drawing
24     JFrame frame = new JFrame();
25     frame.setDefaultCloseOperation(JFrame.EXIT_ON_CLOSE);
26     frame.add(new drawSinewave());
27     frame.setSize(800, 300);
28     frame.setLocation(200, 200);
29     frame.setVisible(true);
30   }
31
32   public static void createSinewave(int Hertz, int volume)
33       throws LineUnavailableException {
34
35     float sampleRate = 44100;
36     byte[] sinesBuffer;
37     sinesBuffer = new byte[1];
38     sinesArray = new double[(int) sampleRate];
39     vol = volume;
40
41     AudioFormat audioFormat;
42     audioFormat = new AudioFormat(sampleRate, 8, 1, true, false);
43
44     SourceDataLine sourceDL = AudioSystem.getSourceDataLine(audioFormat);
45     sourceDL = AudioSystem.getSourceDataLine(audioFormat);
46     sourceDL.open(audioFormat);
47     sourceDL.start();
48
49     for (int i = 0; i < sampleRate; i++) {
50       double angle = (i / sampleRate) * Hertz * 2.0 * Math.PI;
51       sinesBuffer[0] = (byte) (Math.sin(angle) * vol);
52       sourceDL.write(sinesBuffer, 0, 1);
53
54       sinesArray[i] = (double) (Math.sin(angle) * vol);
55     }
56
57     sourceDL.drain();
58     sourceDL.stop();
59     sourceDL.close();
60   }
61
62   protected void paintComponent(Graphics g) {
63     super.paintComponent(g);
64     Graphics2D g2 = (Graphics2D) g;
65     g2.setRenderingHint(RenderingHints.KEY_ANTIALIASING,
66         RenderingHints.VALUE_ANTIALIAS_ON);
```

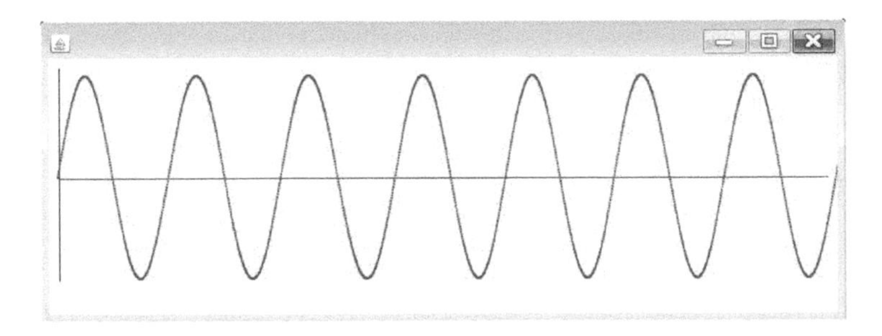

FIGURE 3.23 Drawing sine wave. © Theodor Wyeld. Used with permission.

```
67
68        int pointsToDraw = 4000;
69        double max = sinesArray[0];
70        for (int i = 1; i < pointsToDraw; i++)
71          if (max < sinesArray[i])
72            max = sinesArray[i];
73        int border = 10;
74        int w = getWidth();
75        int h = (2 * border + (int) max);
76
77        double xIncrement = 0.5;
78        // spacing of dots along x-axis
79
80        // Draw x and y axes lines
81        g2.draw(new Line2D.Double(border, border, border, 2 * (max + border)));
82        g2.draw(new Line2D.Double(border, (h - sinesArray[0]), w - border,
83            (h - sinesArray[0])));
84
85        g2.setPaint(Color.red);
86
87        for (int i = 0; i < pointsToDraw; i++) {
88          double x = border + i * xIncrement;
89          double y = (h - sinesArray[i]);
90          g2.fill(new Ellipse2D.Double(x - 2, y - 2, 2, 2));
91        }
92    }
93  }
```

When you compile and run this program you should see something like Figure 3.23. You should also hear a dull tone coming from your speakers. We can create a flow diagram to explain how this works, as can be seen in Figure 3.24.

When we use the rising and falling values of this sine wave to feed our 3D object's animation parameters we will get an idea of what to expect from the full audio file compile.

3.5.2 Basic sine wave-driven animation

In this next build we will use our sine wave generator to provide a rising and falling value to animate our 3D object from the 3D visualiser build (in place

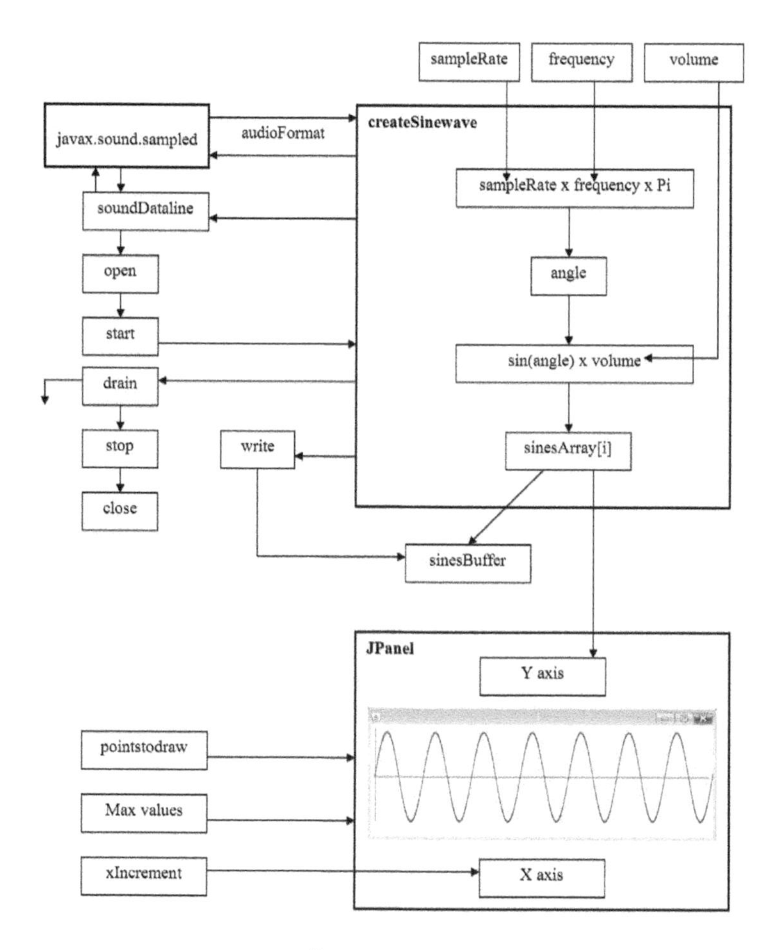

FIGURE 3.24 Flow diagram. © Theodor Wyeld. Used with permission.

of the slider control). But, we will use a trimmed-down version of the 3D visualiser, with the sliders removed to simplify the program (we have added some default values for the transform parameters: Scale, heading, and pitch). In the **main(String[] args)** you will notice we need to use a try with a catch. This is so a dataline can be generated. It opens a path to storing data in the dynamic memory (RAM). Next we specify an **arraylist** of type **double** to store our sine wave values so we can use them later for animating our 3D object. The **createSinewave** method is the same as for the previous build except that we are catching the **sine[i]** values and sending them to the **arraySines** array so we can cast them to the 3D object's inflate-deflate method's rate parameter.

Listing 3.11 Sine wave animation 3D

```
1  import javax.swing.*;
2  import java.awt.*;
3  import java.awt.geom.*;
4  import java.util.List;
5  import java.util.ArrayList;
6  import java.awt.image.BufferedImage;
7  import java.util.Random;
8
9  import javax.sound.sampled.AudioFormat;
10 import javax.sound.sampled.AudioSystem;
11 import javax.sound.sampled.SourceDataLine;
12 import javax.sound.sampled.LineUnavailableException;
13
14 public class sinewaveAnim3D extends JPanel{
15
16   static double[] sines;
17   static int vol;
18   public int Scale = 250;
19   public int Heading = 45;
20   public int Pitch = -26;
21   public static double arraySinesValues = 0.0;
22
23   public static void main(String[] args){
24
25     try {
26       sinewaveAnim3D.createSinewave(200, 100 );
27     } catch (LineUnavailableException lue) {
28       System.out.println(lue);
29     }
30
31     //Frame object for drawing
32     JFrame frame = new JFrame();
33     frame.setDefaultCloseOperation(JFrame.EXIT_ON_CLOSE);
34     frame.add(new sinewaveAnim3D());
35     frame.setSize(500,500);
36     frame.setLocation(200,200);
37     frame.setVisible(true);
38
39   }
40
41   static ArrayList<Double> arraySines = new ArrayList<Double>();
42
43   public static void createSinewave(int Hertz, int volume)
44       throws LineUnavailableException
45     {
46
47       int N = 1; //number of seconds to repeat
48
49       float sampleRate = 44100;
```

```
50          byte[] sinesBuffer;
51          sinesBuffer = new byte[1];
52          sines = new double[(int)sampleRate*N];
53          vol=volume;
54
55          AudioFormat audioFormat;
56          audioFormat = new AudioFormat(sampleRate,8,1,true,false);
57
58          SourceDataLine sourceDL = AudioSystem.getSourceDataLine(audioFormat);
59          sourceDL = AudioSystem.getSourceDataLine(audioFormat);
60          sourceDL.open(audioFormat);
61          sourceDL.start();
62
63          for(int i=0; i<sampleRate*N; i++)
64          {
65            double angle = (i/sampleRate)*Hertz*2.0*Math.PI;
66            sinesBuffer[0]=(byte)(Math.sin(angle)*vol);
67            sourceDL.write(sinesBuffer,0,1);
68
69            sines[i]=(double)(Math.sin(angle)*vol);
70
71          }
72
73          int samplesToDraw = 2000;
74
75          for(int i = 0; i < samplesToDraw; i++) {
76
77            double y = (sines[i]);
78
79            arraySines.add(y);
80
81            arraySinesValues = arraySines.get(i);
82
83          }
84
85          sourceDL.drain();
86          sourceDL.stop();
87          sourceDL.close();
88      }
89
90      public void paintComponent(Graphics g) {
91
92        Graphics2D g2 = (Graphics2D)g;
93
94        g2.setColor(Color.BLACK); //background colour
95        g2.fillRect(0, 0, getWidth(), getHeight());
96
97        List<Triangle> triangles = new ArrayList<>();
98        triangles.add(new Triangle(new Vertex(1, 1, 1, 0).normalize(),
99              new Vertex(-1, -1, 1, 0).normalize(),
100             new Vertex(-1, 1, -1, 0).normalize(),
101             null, Color.WHITE, -1));
102       triangles.add(new Triangle(new Vertex(1, 1, 1, 0).normalize(),
103             new Vertex(-1, -1, 1, 0).normalize(),
104             new Vertex(1, -1, -1, 0).normalize(),
105             null, Color.RED, 1));
106       triangles.add(new Triangle(new Vertex(-1, 1, -1, 0).normalize(),
107             new Vertex(1, -1, -1, 0).normalize(),
108             new Vertex(1, 1, 1, 0).normalize(),
109             null, Color.GREEN, -1));
110       triangles.add(new Triangle(new Vertex(-1, 1, -1, 0).normalize(),
111             new Vertex(1, -1, -1, 0).normalize(),
112             new Vertex(-1, -1, 1, 0).normalize(),
113             null, Color.BLUE, 1));
114
115       // get normal vector of each plane
116       for (Triangle t : triangles) {
```

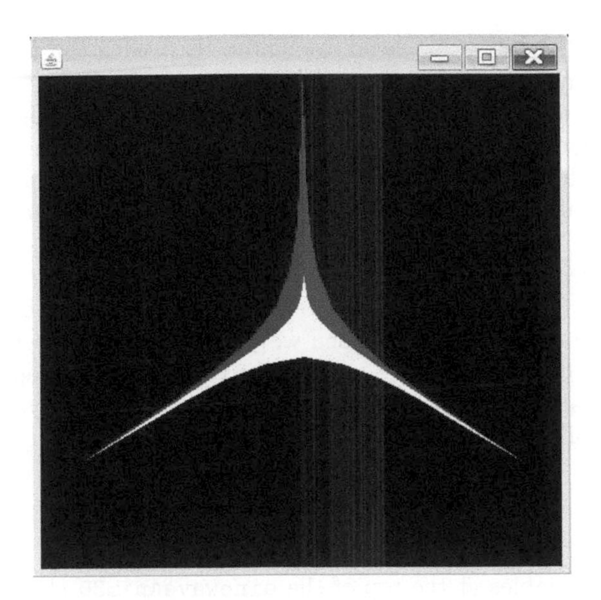

FIGURE 3.25 Sine wave animation 3D. © Theodor Wyeld. Used with permission.

```
117         Vertex normalVector = t.v1.normalVector(t.v2, t.v3);
118         t.normalVector = normalVector;
119      }
120
121
122  ////// inflate method call for spiky or full sphere //////
123
124      double rate = (arraySinesValues + 100)/10;
125
126      for (int i = 0; i < 4; i++) {
127
128         if (rate > 10.0f) {
129           double rate_sub = rate/10;
130           triangles = deflate(triangles, rate_sub, i);
131         } else if (rate <= 10.0f) {
132           double rate_sub = rate;
133           triangles = inflate(triangles, rate, i);
134         }
135      }
136
137  ////// the remainder of this code block is in the appendices //////
```

If you compile and run this code you should see something similar to Figure 3.25.

But there is a problem. We do not see the result until the data has been calculated, and the display only shows the last value. This is because, as in the previous build, the `createSinewave` method needs to execute first before the data is available for drawing the graph (unlike an external file which is already computed and its values are stored). Also, the display only refreshes once – after all the data has been calculated. But with the current build we want our display to refresh in real time so we can see it animate the 3D object. What we need to do is slow down the data transfer rate to something that the display render can handle. To do this, the first thing we need is to create a new thread to handle the data feed to the 3D object (rather than just sending the last value after the **createSinewave** calculations have been done).

First we need to tell the program to redraw the display panel for each update in data. Then create a thread to handle the data which calls the redraw. And finally we need to monitor when we are passing data to the thread and releasing it for display. This uses a **synchronized(this)** statement inside the for loop that takes the **sines[i]** and sends them to the **arraySines** array for access to the 3D object animation rate.

Now add:

```
public static final Object redrawLock = new Object();
```

to the list of variables at the top of the **sinewaveAnim3D** class. And directly under this, add:

Listing 3.12 Redraw

```
1  public void redraw()
2  {
3    while(true)
4    {
5
6      synchronized (redrawLock) {
7        try {
8          redrawLock.wait();
9          repaint();
10       } catch (InterruptedException e) {
11       }
12     }
13   }
14 }
15
16 sinewaveAnim3D()
17 {
18 Thread thread = new Thread(new Runnable() {
19     @Override
20     public void run() {
21       redraw();
22     }
23   });
24
25   thread.start();
26 }
```

Then, inside the

```
for(int i = 0; i < samplesToDraw; i++) { }
```

wrap the following

```
arraySinesValues = arraySines.get(i);
```

with

```
1  synchronized (redrawLock) {
2    redrawLock.notify();
3  }
```

Then, under all of this, add a print line call so we can see what values are being produced

```
1  double values = (arraySinesValues + 100)/10;
2  System.out.println(" values: " + values);
```

Now when you compile and render you hear the tone followed by the printing of a long list of values in the CMD window while the 3D object is animated. But we still have a problem. When the animation is run we get tearing of the video (horizontal splitting of the screen image). This happens when the array of pixels generated by our program is drawn to the screen too quickly – the next array is drawn before the first has completed and so we see a line separating or tearing them. In fact, the only reason we can see the animation at all is because we are printing the values to the CMD. This slows the program down enough that our thread redraw operations can be executed in sequence and the display is updated (if you comment out the println call we only see the end result again). What this tells us is that the data feed rate is too high for the renderer to draw to the screen without overflowing. We need the program to slow down or feed the data at a controlled pace so the renderer can keep up. To do this we need to add a `Thread.sleep()` action. This forces the data feed to pause for a certain period of time (whatever is inside the brackets) before sending the next piece of data.

Under the

```
1  synchronized (redrawLock) {
2    arraySinesValues = arraySines.get(i);
3    redrawLock.notify();
4  }
```

add

```
1  try{
2    Thread.sleep(10);
3    } catch (InterruptedException e) {
4  }
```

This forces the thread to wait 1000/10 ms before sending the next value.

Now when you compile and run the program, after hearing the tone you should see the 3D object animating smoothly and without tearing. Try different `Thread.sleep()` values to see what difference it makes. The following is a flowchart of the main features of our program – which extends the basic sine wave draw program. The full code is available on the book's website.

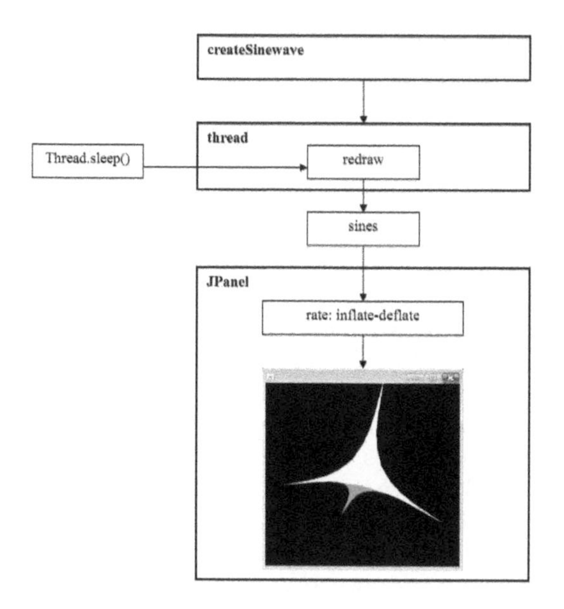

FIGURE 3.26 System flowchart. © Theodor Wyeld. Used with permission.

3.5.3 Full audio file-driven animation

Now that we have the concept of how an audio waveform can be used to drive our 3D animation we are ready to move on to using an external audio file. In some ways this makes our task easier, in other ways it makes it harder. It is easier because the external audio file can be thought of as a container with stored data that we can retrieve (with the sine wave generator we had to generate the data and store it in dynamic memory before we could use it). But it is also harder because there is so much data in an audio file and it does not always generate nice smooth waves – there is a lot of noise which can make our animation seem jittery or cause tearing as discussed earlier.

At the beginning of this section we mentioned how a typical PC can only display information at the rate of about 30 fps. Although 60 and even 100 fps is possible, as Java is a virtual machine – it sits on top of the main operating system (OS) – this means 30 fps is a safe rendering rate. But, an audio file is typically sampled at 44.1kHz. This means it can produce 44,100 samples per second. Hence, just like for the sine wave-generated visualisation, we need to downsample our audio data file. To downsample our audio file to 30 pieces of data per second, we need to know a bit more about it. We need to know how long it is, how many channels (stereo: Left and right), and its bit depth (how many bits per sample).

If our audio file is one minute in length, we need to find the total number of samples. We can do this by simply multiplying 44,100 samples by 60 seconds which gives us a total of 2,646,000 samples. Therefore, if we only want to process our audio at 30 fps we need to get from 44,100 samples per second to 30 samples per second. In other words, we do not want more than 60 seconds of samples at 30 samples per second which equals a total of 1,800 samples. And this means that, in a typical 16-bit audio file with 2 channels (2 bytes per channel = total of 4 bytes per sample), for every 1,470 samples (or 1470 × 4 bytes = 5,880 total bytes) we will produce just 1 frame of values for our 3D animation. Our 1,470 samples are placed in what we call 'buckets' and summed and averaged to give us our 1 frame of data for our 3D animation (each element of the 1,470 pieces of data is added and then the total is divided by 1,470 to get a final average across all the values). This is our 'avg' sample that is sent from the `AudioProcessor` class to the '`Enqueue`' method in the `SpikyTetraSphereAudioDriven` class. We will have 30 buckets of data (with 1,470 avg samples or 5,880 bytes of data in each) for every second of the original audio data (which was at a 44.1khz sample rate).

But, as we are averaging across 1,470 values, this means a lot of data will be lost. It also means there could be big differences between one data value and the next. This would translate into large changes in our 3D animation. For example, if we have a chunk of data A with 1,470 values in it and the next chunk B, also with 1,470 values in it, the chances that the values are close is good only if the original audio file data was something like the sine wave generator – consistently changing over time. If the original audio file contained noisy data (such as for heavy metal[7]) then the differences between chunks of averaged data is more likely to be large.

What we are using is called a low pass filter. It is a coarse method for downsampling which does not result in a very smooth animation. There are better filter algorithms, but they tend to be difficult and quite complex, and introduce their own problems (for example, an FFT filter extracts frequency which does not change as much as the amplitude we are using here). For now, what we are using is easier to understand and provides at least a useful result, and it is suitable for animating from other types of datafeeds (weather, stock market, traffic, etc.) that are perhaps less noisy. But we are not quite there yet.

To work out how many buckets we have, we can take the total length of the audio file (60 secs at 44.1khz) and divide by our bucket size (1,470): (60 x 44,100)/1,470 = 1,800. The data going into the buckets need to be indexed

[7]For example, if the sound file contains heavy metal which has a lot of spurious noise the averaging will not remove all of this and our animation will still look a bit jittery. At the other end of the spectrum, if our sound file contains dub-step style music then the changes in volume and pitch are more gradual, leading to a smoother animation.

(referred to in order). And, our buckets need to be arranged in order too. With our buckets and their contents in order (i.e. indexes(0-28, 29-57, 58-86... etc.)) we can be sure that the animation playback sequence matches the original audio file and the sound coming out of the speakers.

3.5.3.1 Class files

There are three main class files in this build:

- `AudioProcessor.java`,
- `CanvasSpikyTetraSphere.java`
- `SpikyTetraSphereAudioDriven.java`

Each performs a specific function in the overall program. The `AudioProcessor` file processes the audio data and makes it available to the `SpikyTetraSphereAudioDriven` class file. The `CanvasSpikyTetraSphere` class file builds the GUI for the 3D animation.

Just to recap, we are using the output from the audio visualiser in an earlier build to drive the 3D animation (via the inflate-deflate methods – substituting the slider control with a data feed). The main difference from previous audio and 3D builds is that we are using a few different methods – to play the sound, draw it to a 2D graph, and send it to the 3D animation using a queuing process which helps us control how fast the data is being passed on. This build also includes 2 display panels – one for displaying the audio graph as a line waveform in 2D and another for the 3D animation. And finally, we need to establish 4 threads to handle the data: Audio in/out, audio processing, data queuing and the 3D animation visualisation itself. As in the Basic Audio Animated Visualiser build, we will use a 'swing worker' to handle the audio thread in the background. This will significantly improve overall performance of our program.

AudioProcessor class

For the `AudioProcessor` class we have variables at the top, followed by the playback function (`PlayerRef`) which starts the player and displays the audio data on the graph. It takes the loaded file and plays it. It also builds the display content and repaints updates. From here we get the `EventQueue` which launches a dispatch thread to handle the data feed to the 3D object (discussed in more detail later). The audio processor invoker loads the audio file, adds content to the panel and initialises a window for display. The `loadAudio` method takes the input from the audio file in PCM format and iterates it through the channels and adds all the samples to form a single byte array ready to be drawn in the `drawImage` method. It also invokes the 3D object GUI. The Playback loop defines the default 3D parameters, transforms audio bits to bytes, sets timer tasks, and does the averaging, finally sending the audio data in a form that can be used to animate the 2D line waveform display. The Playback loop is done in the background. `displayPanel` does the drawing and launches the main.

spikyTetraSphereAudioDriven

Like most class files, in `spikyTetraSphereAudioDriven` we see the variables are defined at the top. They include default values for what were sliders in previous builds. There are also some new types: A synchronised `redrawLock` object, `keepGoing` Boolean for a `runLoop` method, a static `arrayList<float>`, and `bucketAve` for our data smoothing filter method.

Like the use of synchronised (`pathLock`) in the Basic Audio Animated Visualiser build, in this build a synchronised `redrawLock` object is used to prevent more than one thread accessing the same variable (`bucketAve`) at the same time. What this means is that while two values may be added to the same variable from different sources or threads they are separated so that only one thread has access at a time. In our case, we are creating buckets and then putting data in them. We want the data that goes into the buckets to be placed in the correct order and the buckets to be in the correct sequence. Here is a short program to demonstrate how this works[8]:

Listing 3.13 Synch demo

```
1   import java.util.*;
2
3   public class SynchDemo {
4     public synchronized void test(String name) {
5       for (int i = 0; i < 10; i++) {
6         SynchOutput.print(name + " :: " + i);
7         try {
8           Thread.sleep(500);
9         } catch (Exception e) {
10          SynchOutput.print(e.getMessage());
11        }
12      }
13    }
14
15    public static void main(String[] args) {
16      SynchDemo synchDemo = new SynchDemo();
17      new TestThread("THREAD 1", synchDemo);
18      new TestThread("THREAD 2", synchDemo);
19      new TestThread("THREAD 3", synchDemo);
20    }
21  }
22
23  class SynchOutput {
24    public static void print(String s) {
25      System.out.println(s + "\n");
26    }
27  }
28
```

[8]see: `https://stackoverflow.com/questions/1085709/what-does-synchronized-mean`

```
29  class TestThread extends Thread {
30    String name;
31    SynchDemo synchDemo;
32
33    public TestThread(String name, SynchDemo synchDemo) {
34      this.synchDemo = synchDemo;
35      this.name = name;
36      start();
37    }
38
39    @Override
40    public void run() {
41      synchDemo.test(name);
42    }
43  }
```

When this program is compiled and run you should see a list appear in the CMD window with each new TestThread executed with corresponding iterations from 0 to 9 before the next TestThread is loaded and executed. This is just like our bucket ordering with its contents filled before the next bucket is created and its contents added. This is a form of serialisation.

Now that we have set up a synchronised threading method for our buckets and their content we can implement and send the results to our 3D animation. We do this using a runLoop method. What the runLoop method does is to send the data on in realtime so it updates the 3D animation. To do this we have a while loop which checks to see what state the keepGoing variable is in. Notice that the keepGoing variable is a private volatile Boolean. The Volatile part just means it is held in dynamic memory. In other words, it is deleted or replaced whenever it is not needed. This means the last statement in the loop can be executed after the state for keepGoing has been checked – otherwise it would just stop at the state check waiting for instructions about what to do next.

In short, we have our incoming data thread provided by Enqueue from the AudioProcessor class (discussed in more detail next). We add the values to an array as they come in and are summed and averaged for each bucket. Now, earlier on we said that 44.1khz samples times 60 seconds gives us a total of 2,646,000 samples. But, you will notice right at the beginning of the code, we can set the DEF_BUFFER_SAMPLE_SZ. This needs to be set to the sample size that we need for the downsampling (1,470 x 4bytes per sample = 5,880). This determines the size of data chunks placed in the buffer from our audio file ready for processing. This is where our synchronised thread method works for us. We have two threads trying to fill the same variable (bucketAve) – one thread creating buckets and another thread trying to put content into the buckets. As each bucket is filled, redrawLock() is notified so that our display can be updated in realtime. And, once each bucket is filled, summed and averaged, a new one is created and the process continues (sequentially ordering each bucket's ID and its contents as it goes). To send this on to our 3D animation, we need to create and start the runLoop thread – for passing the data to the 3D animation. Finally, runLoop uses a 'while' loop which

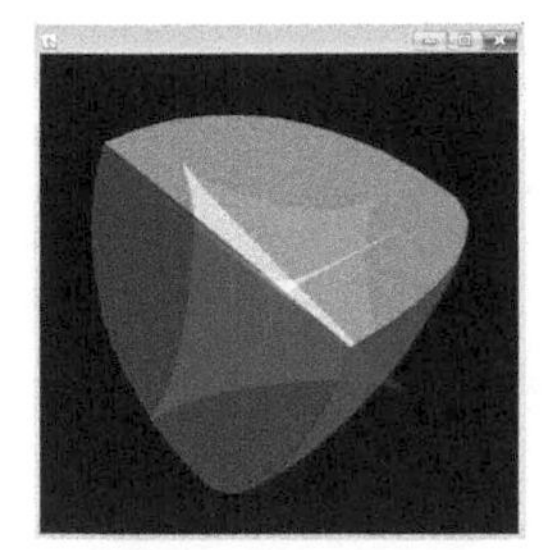

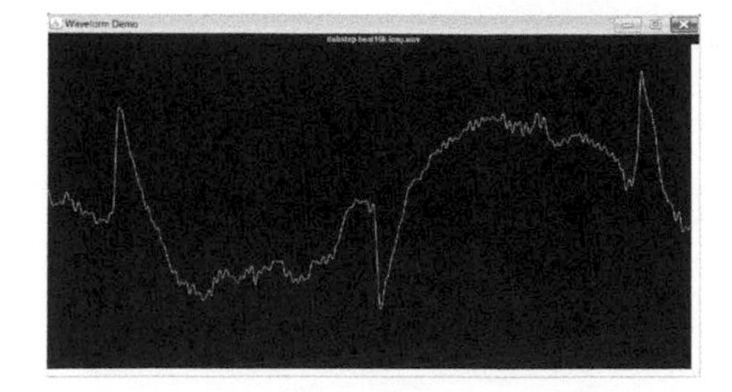

FIGURE 3.27 Spiky tetra sphere. © Theodor Wyeld. Used with permission.

relies on the volatile `keepGoing` Boolean state to re-create itself, synchronise the threads, pass the data on, and call repaint so that our output is displayed in realtime.

Getting back to the role of `Enqueue`: The way enqueue works is, as one value is inserted at one end of an array the first element is output from the other end. The overall size of the array remains the same – size can be set. In other words, when we insert or 'enqueue' we pass on the value or data, and at the other end the data first added is removed or dequeued, returning a value – all the time the array stays the same size. In our case, `AudioProcessor` is sending data via `Enqueue` with an array size of 5,880 bytes. At `spikyTetraSphereAudioDriven` the data is intercepted, summed and averaged and passed to our 3D animation. `Enqueue` provides a constant stream of data, in order. The example below shows how this works.

We can implement a method for finding or setting the `size()` of a data queue and `get()` or return a value to be removed. To do this we can create an array of `size(i)` and then add elements. Every time we add or insert a value into one end of the array (`rear` ++) we can get or delete a value from the other end using (`front` ++). For example:

Listing 3.14 Queue

```
1  import java.util.*;
2
3  public class Queue
4  {
5    int queue[] = new int[5];
6    // create an array of size 5
7    int size;
8    int rear;
9    int front;
10   // could use dequeue to remove elements from front of array
11
12   public void enQueue(int data)
13   {
14     queue[rear] = data;
15     //enqueue or add data to queue array
16     rear = rear + 1; // +1 or ++
```

```
17        size = size + 1; // +1 or ++
18      }
19
20      public void show() // shows the method in action
21      {
22        System.out.println("Elements: ");
23        for (int i = 0; i < size; i++)
24        {
25          System.out.println(queue[i] + " ");
26        }
27      }
28    }
29
30
31    import java.util.*;
32
33    public class Run {
34
35      public static void main(String[] args)
36      {
37        Queue q = new Queue();
38        q.enQueue(5); // add the element 5
39        q.enQueue(2); // add the element 2
40        q.enQueue(3); // add the element 3
41        q.enQueue(7); // add the element 7
42        q.enQueue(1); // add the element 1
43
44        q.show(); // invoke the show method
45
46      }
47    }
```

When this program is compiled and run it returns a list of elements one at a time, and in order. This is the method used to populate our audio array values and control how we pass them to the 3D animation in order.

The rest of the `spikyTetraSphereAudioDriven` class is pretty much the same as for previous builds. When all three class files have been compiled and run you should see something like Figure 3.27. The full code is available on the book's website.

3.6 A FRAMEWORK FOR VISUALISATION OF CODES

3.6.1 How to build a computer and why it's worth knowing

Up to now we have been focusing on building apps using code. We need something to run our code on though. There are lots of different devices we can use to run our code: A mobile phone, a Raspberry Pi, an Arduino board, a MakeyMakey with peripheral devices, or a personal computer. Therefore, when coding, it is important to understand how the computer works, much like understanding how the engine of a car works while driving. Below is a set of visual (see Figures 3.28 and 3.29) and text-based instructions and a self-illustrated pocket book on how to build a computer by University of Northern Colorado student Megan Maddocks. The two pages are intended to be printed twice on the back and front of the same page, then folded horizontally.

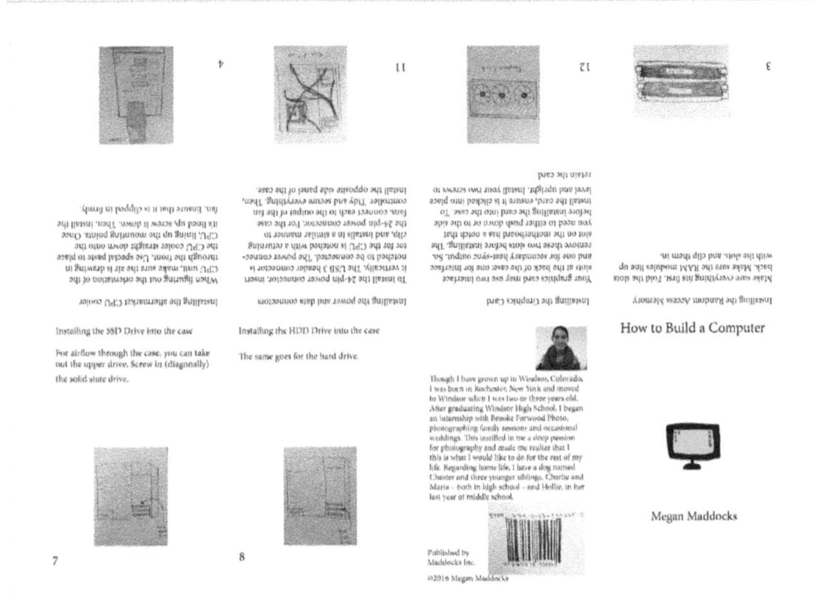

FIGURE 3.28 How to build a computer book, front. © Megan Maddocks. Used with permission.

First, the students designed a book on how to use the computer using their own sketches. Then, they created storyboards for their collaborative videos.

So here is how the computer works, by Megan Maddocks:

1. Anti-Static Protection
   - Use an anti-static wristband. Install the power supply into the case. On the back of the case, there are four mounting points.
2. Installing the CPU (5:50)
   - Arm out, down, and under
3. Installing the RAM (7:10)
   - Make sure everything fits first. Fold the slots back. Make sure the RAM modules line up with the slots, and clip them in.
4. Installing the aftermarket CPU cooler (8:47)

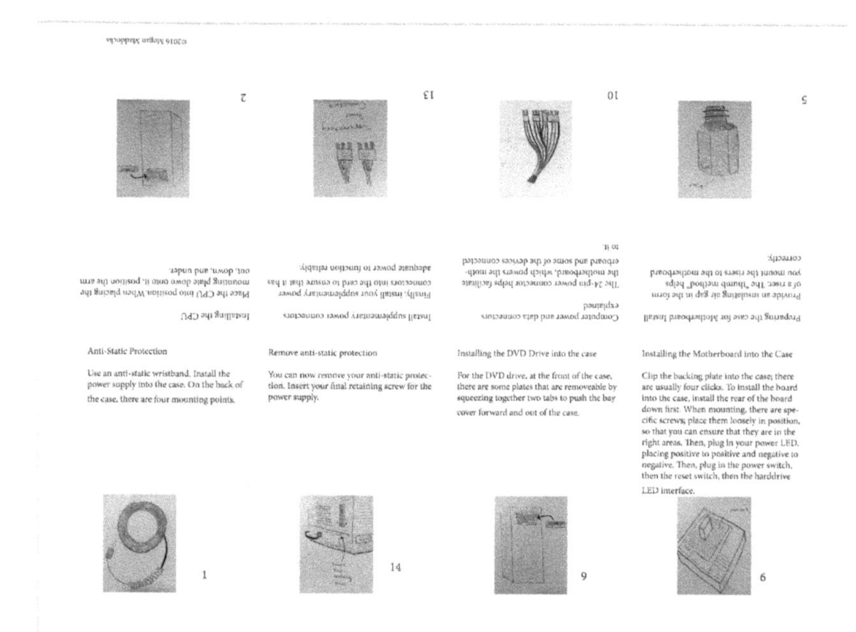

FIGURE 3.29 How to build a computer book, back. © Megan Maddocks. Used with permission.

- When figuring out the orientation of the CPU unit, make sure the air is drawing in through the front. Use special paste to place the CPU cooler straight down onto the CPU, lining up the mounting points. Once it's lined up, screw it down. Then, install the fan. Ensure that it is clipped in firmly.

5. Installing the case for Motherboard Install (13:01)

   - Provide an insulating air gap in the form of a riser. The "thumb method" helps you mount the risers to the motherboard correctly.

6. Installing the Motherboard into the Case (15:06)

   - Clip the backing plate into the case; there are usually four clicks. To install the board into the case, install the rear of the board down first. When mounting, there are specific screws; place them loosely in position, so that you can ensure that they are in the right areas. Then, plug in your power LED, placing positive to positive and negative to negative. Then, plug in the power switch, then the reset switch, then the harddrive LED interface.

7. Installing the HDD, SSD, and DVD drive into the case (18:46)

- For airflow through the case, you can take out the upper drive. Screw in (diagonally) the solid state drive. The same goes for the hard drive. For the DVD drive, at the front of the case, there are some plates that are removable by squeezing together two tabs to push the bay cover forward and out of the case.

8. Computer power and data connectors explained (22:00)

   - The 24-pin power connector helps facilitate the motherboard, which powers the motherboard and some of the devices connected to it.

9. Installing the power and data connectors (23:49)

   - To install the 24-pin power connector, insert it vertically. The USB 3 header connector is notched to be connected. The power connector for the CPU is notched with a returning clip, and installs in a similar manner to the 24-pin power connector. For the case fans, connect each to the output of the fan controller. Tidy and secure everything. Then, install the opposite side panel of the case.

10. Installing the Graphics Card (27:30)

    - Your graphics card may use two interface slots at the back of the case: One for interface and one for secondary heat-sync output. So, remove these two slots before installing. The slot on the motherboard has a notch that you need to either push down or to the side before installing the card into the case. To install the card, ensure it is clicked into place level and upright. Install your two screws to retain the card. Finally, install your supplementary power connectors into the card to ensure that it has adequate power to function reliably.

11. Remove anti-static protection

    - You can now remove your anti-static protection. Insert your final retaining screw for the power supply. Based on: `https://www.youtube.com/watch?v=ObUghCx9iso`

How to make a book, from a traditional to the electronic one, is an important topic in the age of portability and accessibility.

Megan was a project leader for the introductory level class in the Computer Graphics/Digital Media program. Every semester, students received various resolution camera footage of the process of building a computer by two advanced students. They were asked to edit an instructional video as a class based team. The sample videos are posted on the following website: `youtube.com/watch?v=m28186QIsqM` and `vimeo.com/204641130`

3.6.2 Projects involving expressive themes

3.6.2.1 Coding a horse and a rider, across media

Horses have been essential for transportation and other uses for millennia. There are countless functions, associations, and connotations we ascribe to our notions of a horse's usefulness to us. When very young, we might think of a rocking horse, ponies, as well as a knight on a horse flying in the sky, or even a Cinderella with her 'substitute' horses, when her Fairy Godmother transformed a pumpkin and various animals into a carriage with horses, a coachman and a footman. With a saddle or not, circus people perform their acts, and Western cowboys jump on and ride specifically trained horses they had a strong bond with; also, the preparation stage for a horse race may provide opportunities for building individual ways of communication, control, and bonding with a horse. In earlier times, horse-drawn chaises, carts, coaches, stagecoaches, and other carriages served as a means of travelling and for postal deliveries. There are countless instances of horses in art worldwide across time. Greek vases show them around the vases almost offering action, like later on is presented in animations.

From a variety of horse populations, recreational horses provide enjoyment, mental repose, pass time in a healthy way, feel close to nature, make some intellectually involving social bonding, or help gain status in a particular social group. Many horse breeds have been developed for use in sport. Some sports involving horses include hunting, horse racing with betting on favourite horses, playing polo, horse riding, skijoring (Scandinavian tradition of riding on skis pulled by a horse), ancient Greek and Roman chariot races, modern harness racing (mostly with two-wheeled carts), equestrian sports at the summer Olympics, and participating in a kulig (an old Polish tradition of a party involving a cavalcade of horse-pulled sleighs), among other sports. Different seasons often dictate specific sports with the use of various kinds of horses.

Horses differ in terms of their size, colours, origin, and common uses; it is said there are more than 350 horse breeds of riding horses, show horses, sport horses, work horses, ponies; there are the arabian, anglo-arabian, appaloosa, percheron, mustang, and Przewalski wild horse. Horses were even taught to bite people to make them dangerous in battles and in police actions against participants of street riots and rallies, and during wars as well. Over the centuries, there has been severe suffering of horses used in coalmines, and for transporting coal, lumber, or potatoes. Horses served the nomadic peoples both on the prairies of the western states of the USA and eastern steppes of Eurasia. There is a long history of using horses in warfare, in old historic battles (such as those involving winged hussars, whose horse carried long wings to scare the opponent's horses by the sound they produced, in the Kingdom of Poland and other cavalry formations). There is a set of traditions related to horses in China. The Tatars (sometimes spelled Tartars) tenderised their meat for dinner by placing it under the saddle while riding their horses. One may think of the British tradition of riding a horse before breakfast (like the Queen of England did with former president Ronald Reagan). After mastering the performance of a unified team of horse and rider, a horse can also be taught to dance, both at horse shows and in a circus.

A great number of people are involved with horses, as breeders and owners. Jobs in the equestrian industry include veterinarian and veterinary technician, instructor/coach, trainer, groom, jockey, saddler, harness maker, saddle fitter, and equine chiropractor. Other jobs with horses listed include a carriage maker/repair, equine massage therapist, mounted police, equine physiotherapist, equine nutritionist, stable manager, barn hand, pony ride operator, pony rider, pony party organiser, trail guide, guest ranch operator/employee, insurance broker/sales, show judge, show manager, show support staff (stewards), lab technician, website designer, writer, hay dealer, photographer, artist, exercise rider, carriage tour driver, trucker/horse transport, dealer, farm inspector, therapeutic riding instructor/support staff, seamstress/designer, toy designer/manufacturer, software designer/programmer, jump/course designer, breed inspector, artificial insemination specialist, catch rider, and equine dentist.

A horse is often associated with the notion of sheer beauty. According to the French writer Honoré de Balzac, "It is a true saying that there is no more beautiful sight than a frigate in full sail, a galloping horse, or a woman dancing" (Honoré de Balzac, Father Goriot) and also, "A horse! A horse! My kingdom for a horse!" (William Shakespeare, King Richard, III, V, 4). A horse may also enhance one's self-confidence and make somebody magnificent and impressive in appearance; for this reason there are so many portraits of a person with a horse and monuments with someone on horseback. There is also the fantastical world of the unicorn, often presented in Flemish tapestry, and in rich iconography, in ancient (for example, with Greek Myths of the Trojan horse) medieval and romantic tales, paintings, and literature.

There is a lot of horseplay. Also, there are horse idioms, sayings, quips, such as 'Don't look a gift-horse in the mouth,' 'A horse has a large head, let it worry about it,' 'From the horse's mouth,' 'Hold your horses!' or 'Cart before the horse.' A horseshoe can be placed above the entrance to a house for good luck. There are also horse-related games, e.g., the horseshoe tossing game. Hair growing on the mane and tail of the horse can be used to make a paint-brush, calligraphy brush, upholstery, fishing line, fabric called haircloth, wigs, hats, bows of string instruments, and craft objects such as bracelets, necklaces, and barrettes. It is believed horses don't like to be alone in a stable at night; they enjoy the company of a cat or a rooster. There is a culture to cloth a horse for a specific role assigned to that horse, for example from a circus horse to a working horse.

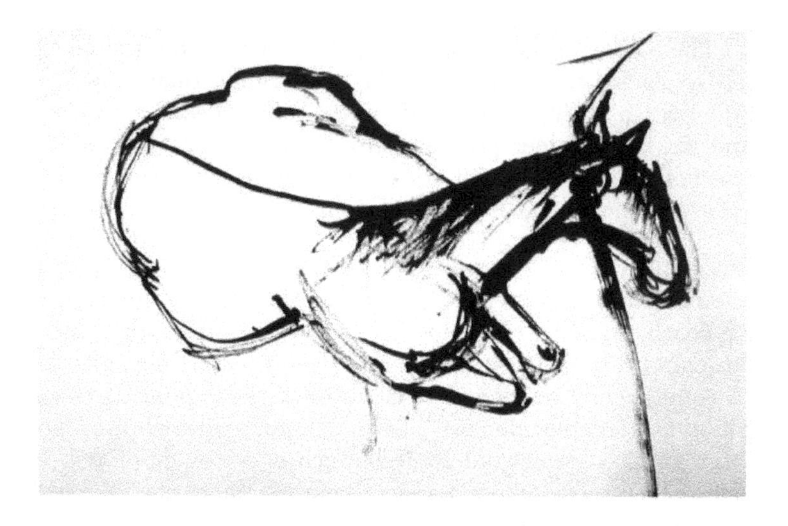

FIGURE 3.30 Ink drawing of a horse. © Anna Ursyn. Used with permission.

The performance by Marion Laval-Jeantet and Benoit Mangin "May the Horse Live in Me" was awarded in 2011 with Golden Nica at Ars Electronica. It was a hybrid bio-art project as part of therapeutic research and an extreme body art where blood plasma of a horse was crossed with a human organism, and then the artist dressed herself in black wearing horse hooves and posed together with a horse.

Since we once relied on horses so much, people created the Pegasus, an imaginary human united with a horse.

Figures 3.30-3.39 show horses created by Anna Ursyn in various media consecutively, from planning (sketches), coding, to transforming the codes to other media.

3.6.3 Scientific visualisation

For creating art we can not only be inspired by our surroundings, but by thinking and analysing processes and products as a source of mental imagery. For example, Figures 3.44, 3.45 and 3.46 show code developed by the University of Northern Colorado student, Sean Flannery, who examined science based events and wrote some code in Processing. The student was asked to visualise scientific concepts in Physics: Acceleration and in Astronomy: The Big Bang Theory and a Black Hole:

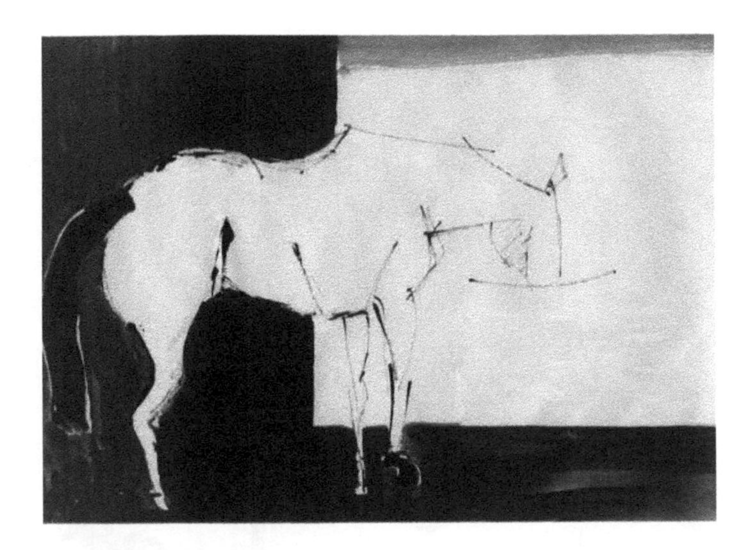

FIGURE 3.31 Ink drawing of a horse on a textured paper. © Anna Ursyn. Used with permission.

3.6.4 Designing computer graphics projects for programming 3D forms

A set of selected nature- and science-inspired learning projects enhanced by computing was designed. Advantages come from the visual/verbal approach to instruction in programming. The visual approach comes from the usefulness of playful ways of learning and also the numerous career opportunities such experiences may open.

A visual solution for a project can be achieved with the use of programming or computer graphics software. The task of completing projects helps students learn to design and write the source code for computer programs as well as create visual content. This approach aims at content analysis linked with a study of techniques and concepts such as independent and dependent variables, with possible comparisons in time and space. A project's theme may inspire solutions and invoke its visual power. Basic art concepts, elements and principles of design can be applied to writing a program for a selected medium.

The reader, apart from learning how to code, should think about their portfolio, writing a résumé, and about the methods and techniques used by companies for hiring. This calls for thinking about the outcome of visual solutions, the use of shortcuts, and including the fun element. For example, if a reader feels tired of coding but still has a good idea for a project, he or she could send parts of their project (separate codes) to a video editing software platform. Codes can then be used in one's resume as a movie file or

FIGURE 3.32 An output from the 3D computer program for a horse altered due to some transformations. © Anna Ursyn. Used with permission.

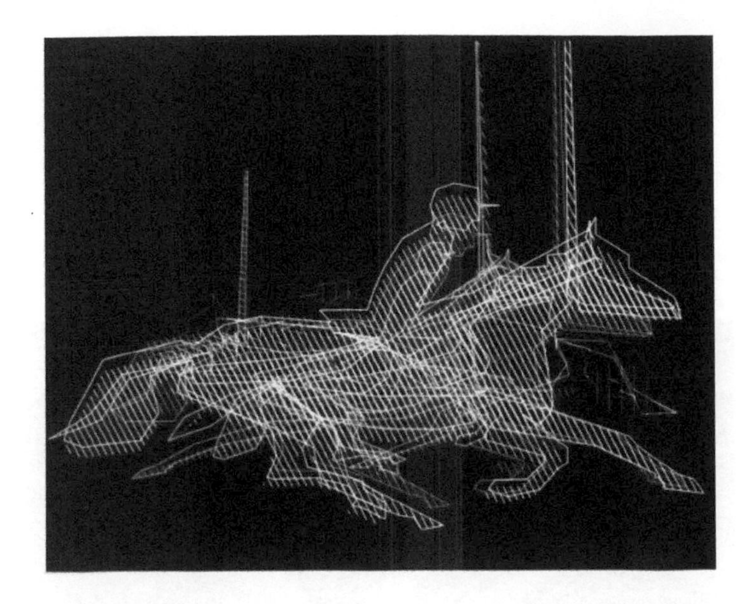

FIGURE 3.33 3D programmed wireframe of a horse created for a colour output. © Anna Ursyn. Used with permission.

as still images. They could be stitched together in a video editing program (see: Video resources for free or paid video edited programs). To prepare a quality portfolio, include an item you are proud of in a visual, permanent, still growing code-based gallery, or on GitHub. In addition to the convenient view of various solutions of projects you will have occasion to discuss them, get advice from fellow students, and provide them with some feedback. For this reason, think about writing a professional critique of a project. Writing a critique means describing, evaluating, comparing, contrasting, rather than just saying personal things about it. Critical analysis is considered a higher level thinking strategy which involves description, analysis, interpretation and judgment.

When we are talking about or writing about art, we use several types of questions to evaluate what we see and what we experience:

- **The descriptive questions** ask what we see;
- **The analytical questions** ask what are the interrelationships of the parts of the artwork;
- **The interpretative questions** ask what the meaning is to you; and
- **The evaluative questions** ask whether it is good or bad art, and whether you like or dislike it.

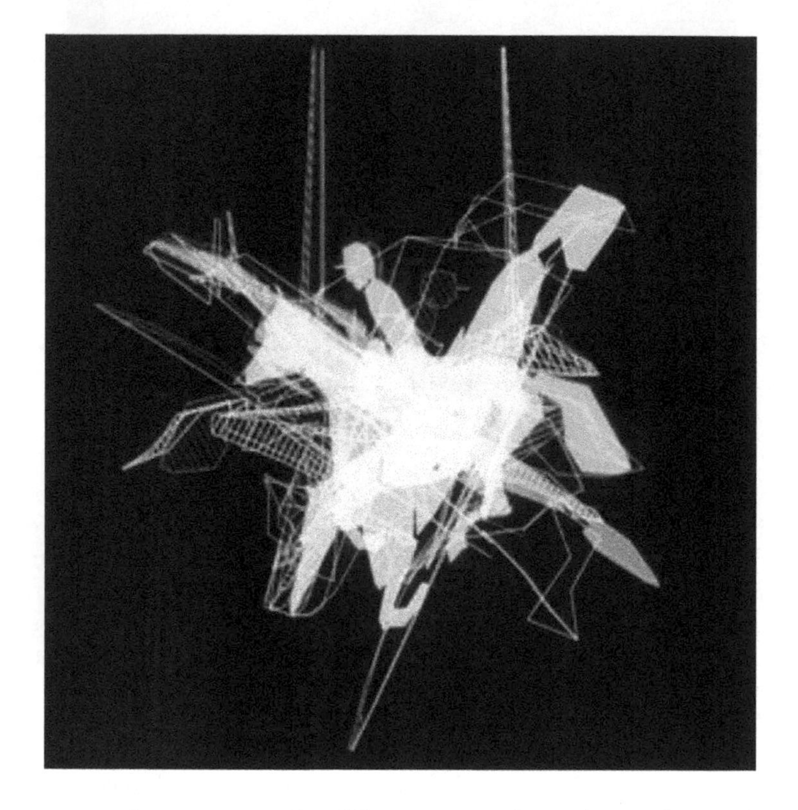

FIGURE 3.34 An output of a 3D program for a horse layered, transformed, and overlapped for a colour output. © Anna Ursyn. Used with permission.

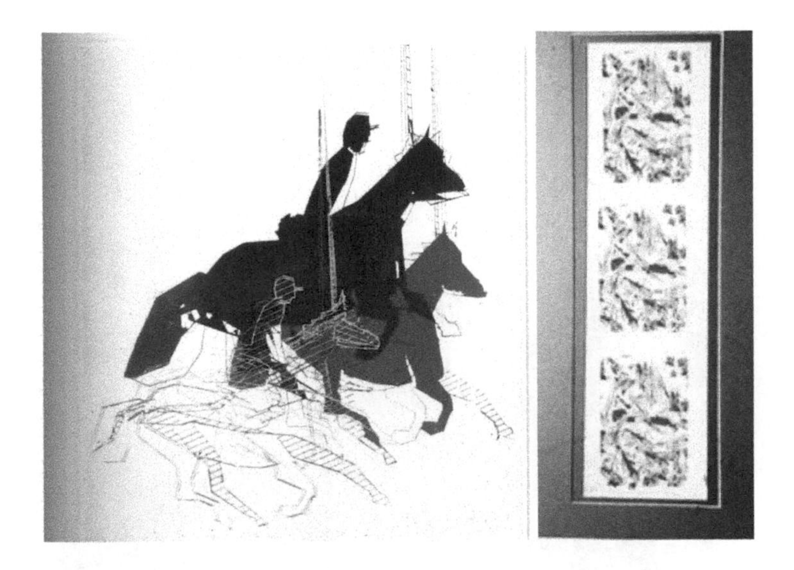

FIGURE 3.35 Photosilkscreens after a black and white programmed horse. © Anna Ursyn. Used with permission.

As your quality portfolio comprises art products, projects may be designed as digital collages that mix programming with software for creating visual products. Projects may be colour coded or designed as grayscale (black-and-white) digital images where the value of each pixel, varying from black to white, carries intensity information [23].

3.6.4.1 Creating characters with feelings

When going from computer graphic based art exercises to the programming, we can design assignments about relations between our senses (such as sight, hearing, touch and haptic experience, smell, taste, and many more) and feelings. This exercise will facilitate the concept of conditions in writing programs that tell how particular senses respond to specific changes in the environment, and what emotions they may evoke. Changes in environment may, for example, include visual, acoustic, chemical, tactile, electrical, and seismic signals.

Design an imaginary character set in a changing environment, involving many senses. Connect your character's perception with changes in its surroundings. The program may determine responses of a particular sense to a change in environment. For example, a change of colour will evoke a response from the character's sight; noise or music will stimulate its hearing; the presence of a rose or cheese will stimulate the smell receptors in its nose; an apricot, when grabbed, will trigger a response from the sense of touch; drinking lemon

FIGURE 3.36 A photosilkscreen printed on acetate after a programmed, then transformed image of a horse. © Anna Ursyn. Used with permission.

FIGURE 3.37 Layered photolithographs after coded images of a horse on an airbrushed surface. © Anna Ursyn. Used with permission.

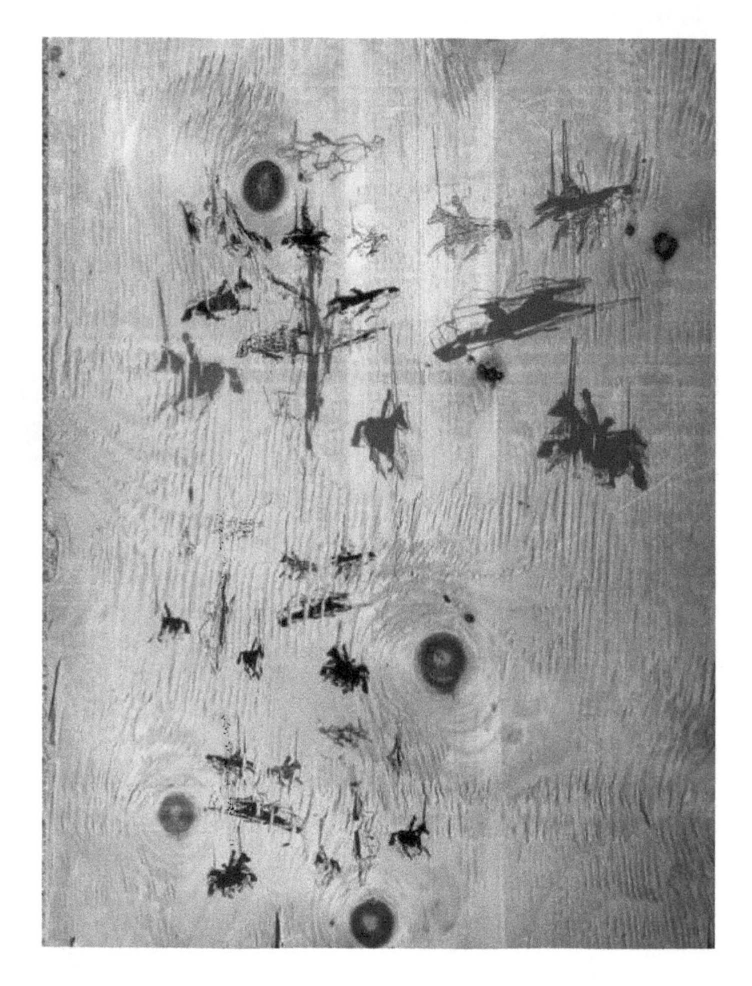

FIGURE 3.38 Photosilkscreen after coded image printed on a wooden plank. © Anna Ursyn. Used with permission.

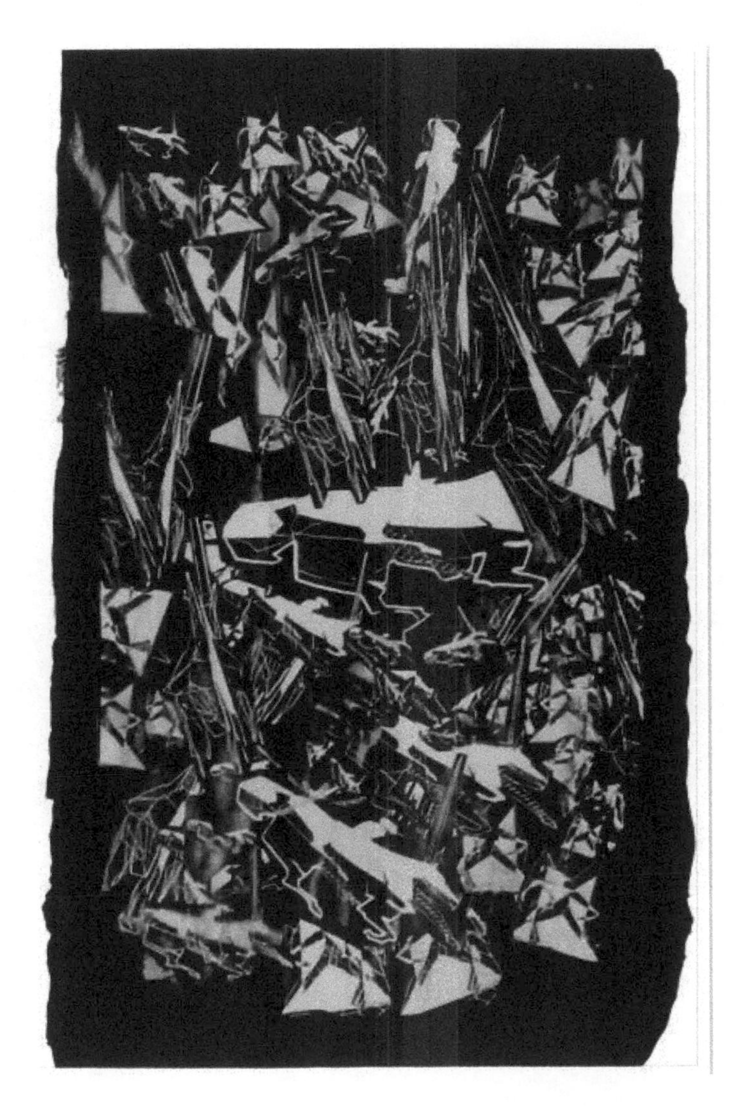

FIGURE 3.39 Photolithograph after various coded plots. © Anna Ursyn. Used with permission.

FIGURE 3.40 A sculpture of a horse with coded images in black and white and in colour for its shadow. © Anna Ursyn. Used with permission.

FIGURE 3.41 "Hero Horse". A sculpture of a horse with several coded images, on glass, wood and metal, with a casted horse and welded frame. © Anna Ursyn. Used with permission.

FIGURE 3.42 “Speeding”. A wooden sculpture of running horses with wood cut according to the output of the codes for the transformed horses and their riders. © Anna Ursyn. Used with permission.

FIGURE 3.43 "Confusion". A sculpture of running horses with the photolithographed images of various horses following their respective codes. © Anna Ursyn. Used with permission.

FIGURE 3.44 Acceleration. © Sean Flannery. Used with permission.

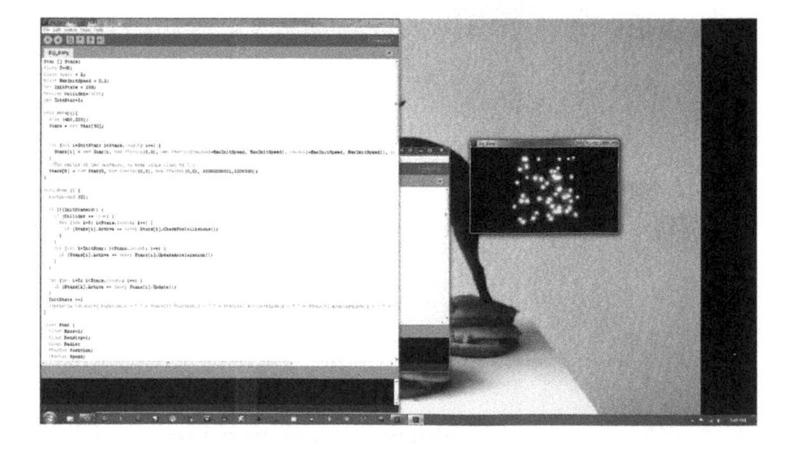

FIGURE 3.45 Big bang theory. © Sean Flannery. Used with permission.

FIGURE 3.46 Black hole. © Sean Flannery. Used with permission.

juice or licking honey will cause a response from its taste buds. An earth tremor may stimulate its internal mechanoreceptors or geospatial sense of the direction.

Changes to our surroundings can be good or bad for us. We may also perceive them as good or bad (we may like hot air when we are feeling cold or we may avoid it when we feel overheated, such as in the condition of hyperthermia). We may thus act at a physical level (for example, a hot surface can burn our skin), physiological level (e.g., climbing at a high elevation can cause a change in our blood level), and psychological level (for example, by evoking emotions such as 'I like it', 'I don't like it', 'I'm afraid of it'). Emotional reactions can act as learned, cliché responses (e.g., colour blue can mean sadness), or may result from complex reactions coming from various brain structures.

From this we can devise a program about conflicting emotional reactions to perceptions coming from our senses, where one's wish, inclination, or desire is conflicting with fear, a dislike, or an avoidance reaction. For example, one can know that sunbathing can cause skin cancer but one may love tanning; one may realise that drinking a lot of coffee (that was not decaffeinated) may advance osteoporosis by lowering the bone calcium level but one may like the taste and smell of coffee and have a good feeling coming from the rise in blood pressure and giving us more energy. Many companies are working on platforms delivering multi-sensory experiences for gaming, health, fitness, hand-eye coordination, biofeedback or behavioural changes, such as habits, or addictions. Some are focusing on entertainment, using functions and tools such as GPS or an accelerometer in order to deliver experience. For example, when it rains where the player lives, it might rain inside the game too.

FIGURE 3.47 Characters and their moods. © Kyle Hathcoat. Used with permission.

This may become more complicated when we try to devise a program about how these good, negative, or conflicting emotions can evoke a feeling of self-indulgence in our minds, controlled by brain structures and mediators. Even more complex would be descriptions of how memories emerge, based on sensual perceptions and the resulting emotions. We may start from activity: Take pictures of your face showing different emotions, and then draw these emotions using apps installed on your phone.

Figure 3.47 is an example of characters portraying their moods, developed by the University of Northern Colorado student, Kyle Hathcoat.

3.6.4.2 Storytelling: Programs for visual/verbal interactions

In this task, with the use of coding, develop visual presentations of various kinds of data. Composition of the project can start with choosing a character and its environment, followed by thinking about the recipient of the project. Every audience has its own needs. Typography – setting types and the appearance of printed material – makes the written part of the project legible, readable, and appealing for the given recipients. For example, small type is not good for small children and the elderly, while a big font size may be seen offensive for high school students.

Create your own visual story and add it to a database for others to critique. Projects show, by the use of coding, individually created environments and reactions of invented characters (avatars) to changes in particular environments. First, choose a theme, a subject that is being dealt with, and objects to which actions will be directed. Then, understanding is needed of underlying processes, products, and forces defining the concept development. Each project starts with choosing a character. That means research is needed about characters and their habitat, and also factors such as symbiosis, dangers, characters' equipment and implements (is your character big, fast, strong, or wise?).

Then comes thinking about who is the audience. Create your storyline – the narrative of your fictional or true research-based story, and then consider possible ways of redesigning this storyline for a particular audience. Write your storyline the way you would tell it on a cell phone; it should be short and interesting.

The story can be retold depending on the media: A graphic, a graphic novel, a comic, manga, animation, theatre performance, VR, a game, a movie, a radio show, webcomic, interactive storytelling, etc. Choosing a media container for a story will define how the story is being delivered in a setting appropriate for telling the same story in another medium. A particular setting may be appropriate for re-telling your story. For example, you might write a short story, and rework it in several ways, thus creating various kinds of containers for the same story according to specific settings and styles, and then visualise your writing and assign your story some visual containers by creating illustrations or animations. Avatars can convey their stories through music, dance, and text. A program or computer graphics may serve for defining a container for a story. Old, popular tales are retold that way in various media.

Try writing a storyline in a way you would tell to your friend over your mobile phone. Make it short, interesting, without visual cues. You may write a story that follows storytelling techniques (exposition, rising action, conflict, climax, falling action, resolution, etc.) and utilises themes, then draw art from there. After the storyline is ready, make a storyboard that organises graphically your storyline with the use of illustrations or simple drawings. Storyboards are used for pre-visualisation of your video, animation, motion picture, motion graphic, or any interactive medium.

After your storyline is ready, create a storyboard – a graphic organiser, a set of drawings or sketches going together with dialogue and directions. It shows the shots you will make for your production. 2D or 3D storyboard software and free printable storyboard templates are available on the Internet[9].

Pick a template that meets your aesthetic and intellectual needs. Select your keyframes defining the starting point and the ending point for smooth transitions between images. A set of key frames will define the movements seen by a viewer. Insert your major key frames into the template and write captions under every change that will explain each dynamic action.

Figures 3.49 and 3.50 present University of Northern Colorado students: Galt Tomassino's work on a game, and a Storyboard for an animation about an insomniac by Christian Eggers. One may also consider making mental keyframes in the programming process.

[9]For example, 3D Rapid Storyboarding is presented here: `https://www.youtube.com/watch?v=qQwatDXGttM`

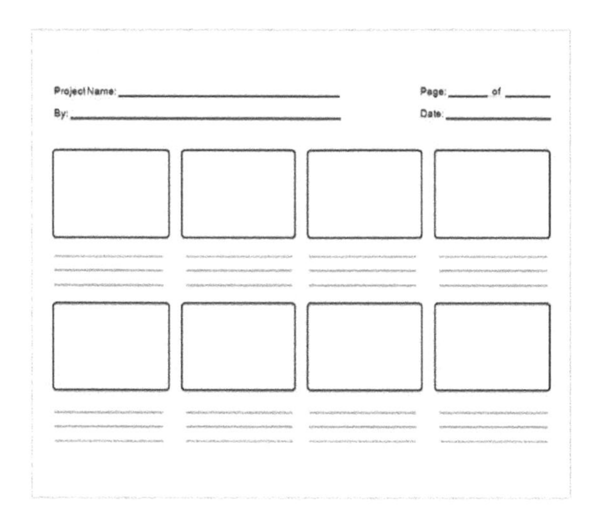

FIGURE 3.48 An example of a template for a storyboard.

Make sketches for your story. Familiarize yourself with sketches made by artists such as Danny Coyeman, who has a habit of making a memory sketch of a cashier giving him his receipt on the back of that receipt and Ken Bernstein, who doodled so hard that the City of Boulder used his doodle as a decoration on the cement wall dividing habitable sections from noise and dust of the busy 28th Street. Make some drawings such as of a mannequin, some geometric sketches, and explore by sketching, and observing the positive–negative space relations.

Sketching has cognitive power, enhances the eye–hand–brain coordination, as well as the mind–brain–eye–hand coordination. The application of sketching is often used in storyboarding, brainstorming, for designing cognitive maps, making web trees, planning events and competitions, or arranging conflict reductions. Dan Roam's book "The Back of the Napkin" discusses such cases [43].

Read about the elements and principles of design in arts, as well as in other disciplines such as coding, written by one of the authors and students of each author. Write your own statement on this theme.

Your drawing or sketches will serve as frames for your visual storytelling. In video technology a frame is a coded image. If you would like to create an animation, you can also define your key frames, which will define starting and ending points or transitions for the timing of the movement. The remaining frames will be filled as inbetweens or 'tweens'. Have your ears open for interesting phrases or conversations as inspiration.

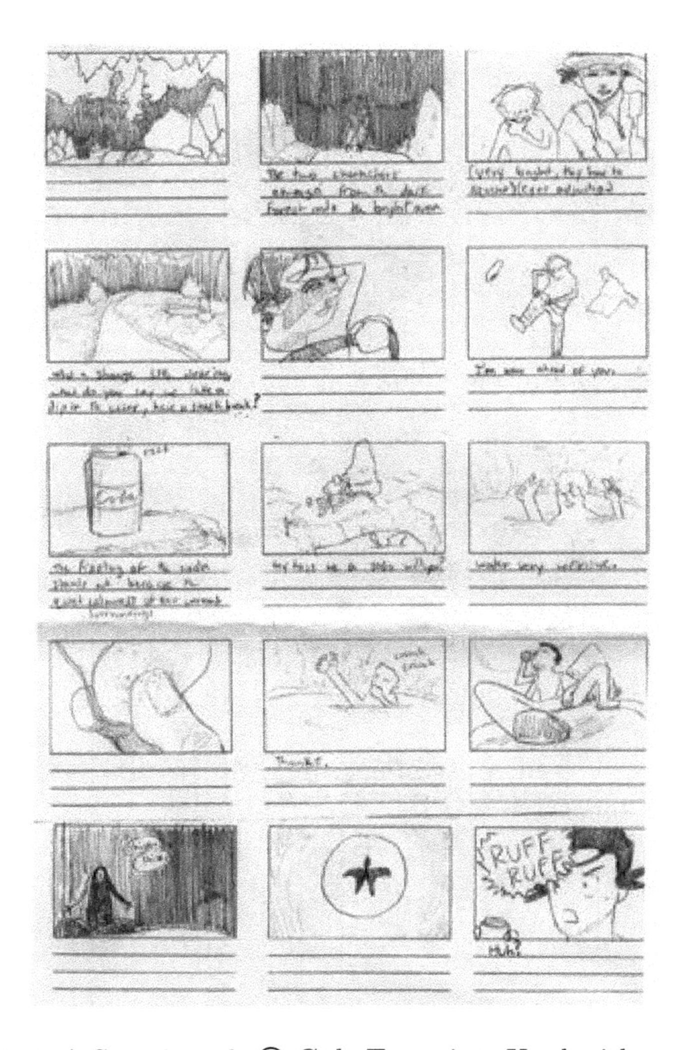

FIGURE 3.49 A Storyboard. © Galt Tomasino. Used with permission.

FIGURE 3.50 A Storyboard. © Christian Eggers. Used with permission.

CHAPTER 4

Interactivity and Visualising Inputs: Mouse, Data, Finger

CONTENTS

4.1 INPUTS AND OUTPUTS

So far we have been making art using code without paying too much attention to how it is viewed or how we interact with it. In this next section, we will investigate using a mouse to guide a graphic on the screen, use a datafeed to animate a graphical tree, and build a physical device with buttons for lighting up an array of LEDs. It is hoped this will inspire you to try your own versions with uniquely creative results.

4.2 A PORTRAIT: PERSON WITH A JETPACK

By Matt Anderson
University of Northern Colorado

This project is written in Processing. The Processing environment can be downloaded from `processing.org`.

Below is code created by Matt Anderson, the University of Northern Colorado student, showing a person with a jetpack created with Processing.org

Listing 4.1 Person with a jetpack

```
1  PImage jetpack;
2  PImage mannequin;
3  JetFuel stream;
4  boolean clickEvent = false;
5  boolean jetOneClicked = false;
6  boolean jetTwoClicked = false;
7  boolean strapClicked = false;
8  boolean mannequinVisible = false;
9
10 void setup(){
11   size(800,800);
12   jetpack = loadImage("Jetpack.png");
13   mannequin = loadImage("mannequin.png");
14   image(jetpack,249,100);
15   stream = new JetFuel();
16 }
17
18 void draw(){
19
20   if ((clickEvent && jetOneClicked)||(clickEvent && jetTwoClicked)){
21     for (int x = 0; x < 25; x++){
22       stream.addParticle(new PVector(mouseX,mouseY));
23     }
24       clickEvent = false;
25       jetOneClicked = false;
26       jetTwoClicked = false;
27   }
28   stream.run();
29   if(clickEvent && strapClicked){
30     if(!mannequinVisible){
31     scale(0.75);
32     image(mannequin, 315, 100);
33     mannequinVisible = true;
```

```
34        }
35        else{
36          translate(20,20);
37          image(mannequin, 315, 100);
38        }
39        strapClicked = false;
40      }
41    }
42
43    void mousePressed(){
44      clickEvent = true;
45      if (mouseX >= 249 && mouseX <= 341 && mouseY >= 335 && mouseY <= 379){
46        jetOneClicked = true;
47      }
48      if (mouseX >= 459 && mouseX <= 551 && mouseY >= 335 && mouseY <= 379){
49        jetTwoClicked = true;
50      }
51      if(mouseX >= 341 && mouseX <= 459 && mouseY >= 100 && mouseY < 235){
52        strapClicked = true;
53
54      }
55    }
56
57    // An array to manage particles
58    class JetFuel {
59      ArrayList<Particle> particles;
60      PVector origin;
61
62      //Constructor
63      JetFuel() {
64        particles = new ArrayList<Particle>();
65      }
66
67      void addParticle(PVector position) {
68        origin = position.copy();
69        particles.add(new Particle(origin));
70      }
71
72      void run() {
73        for (int i = particles.size()-1; i >= 0; i--) {
74          Particle p = particles.get(i);
75          p.run();
76          if (p.isDead()) {
77            particles.remove(i);
78          }
79        }
80      }
81    }
82
83
84    // A simple Particle class
85    class Particle {
86      PVector position;
87      PVector velocity;
88      PVector acceleration;
89      float lifespan;
90
91      Particle(PVector l) {
92        acceleration = new PVector(0, 0.05);
93        velocity = new PVector(random(-1, 1), random(-2, 0));
94        position = l.copy();
95        lifespan = 150.0;
96      }
97
98      void run() {
99        update();
100       display();
```

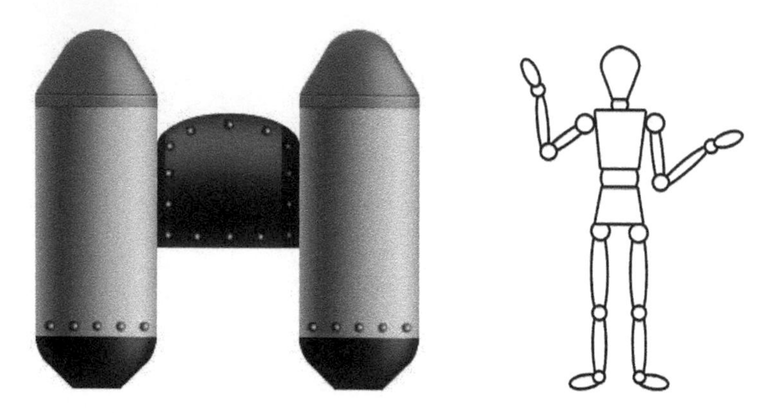

FIGURE 4.1 Reference images. © Matt Anderson. Used with permission.

```
101     }
102
103     // Method to update position
104     void update() {
105       velocity.add(acceleration);
106       position.add(velocity);
107       lifespan -= 1.0;
108     }
109
110     // Method to display
111     void display() {
112       stroke(255, lifespan);
113       fill(255, lifespan);
114       ellipse(position.x, position.y, 8, 8);
115     }
116
117     // Is the particle still useful?
118     boolean isDead() {
119       if (lifespan < 0.0) {
120         return true;
121       } else {
122         return false;
123       }
124     }
125   }
```

The images in Figure 4.1 represent some reference images.

4.3 INTERACTIVE WEATHER TREE APPLICATION

THIS part can be seen as an introduction to interactivity. In order to understand how graphics can be combined with an external data feed (like the latest news, wiki posts, weather and so on) the following project uses a datafeed to update and animate a graphical user interface (GUI). In it you will use Java to create an animated image of a tree which is driven by some external realtime weather data.

There are five class files in this 'weathertree' application: Canvas, **ControlFrame**, **OpenWeather**, **Rain** and **Tree**. Each handles a different function of the overall application. The main file is **Canvas**. This is the reference file that the others get called from. The process includes creating classes containing their own definitions, declaring variables (local inside a method, and instances outside a method), defining methods (functions), and executing the application using a **main()** method. The following explains how each class works.

4.3.1 Basic Tree class

To create a class called '**Tree**' all we need to do is declare the class name and provide it with some curly braces to contain the functionality of the class – its methods:

```
class Tree{ }
```

Next we need to create some instance variables for the parts of our tree that we want to be able to change:

```
1  int x = 300;//declared instance variable
2  //type is int[eger], name is x, initial value =[assignment] 300
3  int y = 350;
4  int branchingNum = 7;
5  int length = 85;
6  int width = 25;
7  int angle = 30;
8  int leafRandoms [];
9  Color treeColor = new Color(142, 98, 89);
10 Color leafColor = new Color(62, 142, 73);
11 GradientPaint backgroundColor = new GradientPaint(0, 480, Color.BLACK, 0,
       240, Color.WHITE);
```

The **int** refers to integer. Other types of numerical values we could define include **double** (0.1) and **float** (0.001). Variables should always be assigned an initial value (use of the = symbol here means to assign the following value to what comes before the = symbol. It does not mean 'equals to'. If you need something to be equal to something else you need to use ==). If the variable is a class it can be given the initial value null; 0 for numeric variables; '\0' (ASCII value for 0) for characters; and, false for Booleans. Variable names can start with a letter, an underscore (_), or a dollar sign ($), but they cannot start with a number. After the first character, they can include any letter or number. Symbols, such as %, *, @, and so on, may be reserved for operators in Java, so symbols are best avoided. Most operations in Java are case sensitive; hence two variables with the same name such as Background and background refer to two different instance variables. It is usual to use **camelFont** when naming variables, classes and methods. Variables always include a type. These include any of the eight basic primitive data types, class names or an array. The primitive types include: Integers (**byte**, **short**, **int**, **long**), **float** (32 bits), **double** (64 bits), **char**/**string** (text), **boolean** (true or false).

Next we want to add a change in behaviour for our class. We do this by defining a method. For example, we can define the background colour of our canvas:

```
1  void paintBackground(Graphics2D g) {
2    // Void means to override the method and replace it with exactly what we
       want
3    // In this case Paint our backGroundColor to a Rect[angle] 640 by 480
       pixels
4    g.setPaint(backgroundColor);
5    g.fillRect(0, 0, 640, 480);
6  }
```

This method calls on the **Graphics2D** class which is an extension to the Graphics class that comes bundled with the JDK. The Graphics class includes methods for creating lines, shapes, characters and displaying images. It includes the standard primitive shapes: Lines, rectangles, polygons, ovals and arcs. To draw an object the **paint()** method in the Graphics class is called. The **paint()** method is used to display graphics initially and when the window is moved, resized or another window is moved over it. The method **repaint()** is used when graphics components are updated – such as in an animation or when affected by an event such as a slider movement or other controller. Note that **update()** is called first. This clears the screen and shows only the background colour before repainting the graphics elements. This can cause flickering. To avoid this we can override **update()** by specifying our own version which does not update the part of the paint method which clears the screen. Instead it just gets redrawn. Alternatively we can just redraw what needs to be redrawn instead of the whole screen using clipping. It involves tracking graphics objects as they move across the screen and only updating those parts that need to be updated. This is a complex concept so it is not discussed here. At least you will know what to search for. Flicker can also occur when using images in an animation. Another method for avoiding flicker is to create a copy of the graphic elements offscreen where the painting is done. Once complete the whole image is used to update the main screen. This is called double-buffering. It involves effectively creating a dummy screen that does not get seen but is called as an instance variable so it can be passed to the **paint()** method. At the end of the **paint()** method the offscreen buffer gets drawn to the real screen.

All drawing methods have arguments representing start, end and corner locations for the object to be drawn. The origin (0,0) for the coordinate system in Java is in the top left corner. Positive x values go to the right, and positive y values go down. Measurements are in pixels which are always represented as whole integers. For example, if we want to draw a rectangle we can use either the **drawRect()** or **fillRect()** methods of the **Graphics** class. The first draws a rectangular outline, the latter fills it.

```
1  public void paint(Graphics g) {
2    g.drawRect(10,10,50,50);
3    g.fillRect(70,10,50,50);
4  }
```

In this case our outline rectangle starts at x = 10, y = 10 and has a height and width of 50 pixels each. Our filled rectangle starts to the right of the outline rectangle at x = 70, y = 70 with a height and width also of 50 pixels each. We can round the edges of our rectangles by using the **drawRoundRect()** or **fillRoundRect()** methods instead. They include an extra two arguments for the width and height of the arc at the corners. Similarly, we can the use the **drawPolygon()** or **fillPolygon()** methods to construct more complex shapes. The extra arguments refer to the x and y coordinate for each vertex of the polygon. Ovals can be created using the **drawOval()** or **fillOval()** methods. They each use only four arguments. The first two refer to the x and y starting coordinate; the second two arguments refer to the eccentricity of the oval. If the two values are the same a circle is drawn. If they are different an ellipse is drawn. Arcs are drawn slightly differently. We can use the **drawArc()** or **fillArc()** methods. Each contains six arguments. The first four arguments refer to the starting coordinate and the height and width of a circle or oval. The last two arguments refer to the start angle in degrees for an arc and how far it is swept in a positive direction.

```
1  public void paint(Graphics g) {
2    g.drawArc(10,10,50,50,90,180);
3    g.fillArc(70,10,50,50,90,180);
4  }
```

In this case the first arc would sweep out a line shape and orientation of a capital 'C'. The second arc would create the same shape but is filled like a half moon. Elliptical arcs can be generated in the same way as for oval shapes. A negative sweep angle could be used to create a mirrored 'C', and so on.

Colours are accessed from the **Color** class. Java uses a 24-bit colour scheme represented by combining red, green and blue values in the range 0 to 255 or explicitly such as the system colours using **Color.red** and so on.

Because we are simply creating a background for our graphic and not drawing shapes we can use the **Graphics2D** extension to the Graphics class to fill our rectangular panel. A method of the **Graphics2D** class is **setPaint** and `fillRect` (fill rectangle). In `setPaint` we are using the initial values specified in our instance variable for `backgroundColor` (black and white). In `fillRect` we have specified a rectangle that is filled from the top left corner and is 640 pixels in width by 480 pixels in height. Our `backgroundColor` specifies that black should start at 480 pixels down the frame in the y direction (at the bottom of the panel) and white starts at 240 pixels in the y direction (halfway down the panel). `GradientPaint` is a subclass of the `Graphics2D` class which defines how the black and white colours behave. In this instance they create a colour gradient or ramp. The standard Java specification includes a lot of predefined colours (red, green, blue, magenta, yellow, orange and so on). Alternatively we can mix our own colours. This is what we have done with the `treeColor` and `leafColor`. The three numbers inside the parentheses next to `new Color` refer to the percentage of red, green and blue colours to mix (the RGB colours). If all three numbers were zeros this would create the colour black. If all three numbers were 255 this would create the colour white. We can test this by using `treeColor` and `leafColor` in place of `BLACK` and `WHITE` in our `backgroundColor` method. But first we need to construct a panel to display our canvas on. The following creates a frame which holds a panel which displays our background colour ramp.

Listing 4.2 Background colour

```
1  import java.awt.Color;
2  import java.awt.GradientPaint;
3  import java.awt.Graphics2D;
4  import java.awt.Graphics;
5  import java.awt.*;
6  import javax.swing.*;
7
8  public class backgroundColor { // class definition
9
10   public static void main(String[] args) { // main() method declaration
11
12     JFrame frame = new JFrame();
13     Container pane = frame.getContentPane();
14
15     frame.setDefaultCloseOperation(WindowConstants.EXIT_ON_CLOSE);
16     frame.setSize(640, 480);
17     frame.setVisible(true);
18
19     JPanel MyBackground = new JPanel() {
20       public void paintComponent(Graphics g) {
21         Color treeColor = new Color(142, 98, 89);// creates a new
22                                     // instance of Color
23                                     // and
24         // stores it in the variable treeColor
25         // the three values refer to the amount of the //colours red,
26         // green and blue (0 to 255)
27         Color leafColor = new Color(62, 142, 73);
28         GradientPaint backgroundColor = new GradientPaint(0, 480,
29             leafColor, 0, 240, treeColor);
30         Graphics2D g2D = (Graphics2D) g;
31         g2D.setPaint(backgroundColor);
32         g2D.fillRect(0, 0, getWidth(), getHeight());
33
34       }
35     };
```

FIGURE 4.2 `backgroundColor.class` executed. Notice the blending of `leafColor` and `treeColor`. © Theodor Wyeld. Used with permission.

```
36
37         pane.add(MyBackground);
38
39     }
40 }
```

If you save the above code in a text file called `backgroundColor.java`, compile and run it, you should see the screen in Figure 4.2.

Although we are only creating a panel with a background gradient colour or ramp there is a lot going on in this file. At the top of the list we see '`import`'. This is where the standard libraries that come bundled with the JDK are called up. They need to be declared at the beginning so that when they are referred to later in our class structure their instance variables can be adjusted to create our display. A method is used in our main `backgroundColor` class: `public void paintComponent`. The `paintComponent` method is declared public so it can be accessed by other methods and classes. It means it can be accessed from other classes either within the main class or outside of it. A private class can only be accessed from within the main class (more on private classes later). For example, our main `backgroundColor` class is responsible for creating a frame (`JFrame`) to contain anything else we want to happen in our class. In this case we want to create a panel to display our background colour ramp. This is why the panel part is `public`, so it can be accessed from the main class. The `JFrame` class is really just a container: `Container pane = frame.getContentPane();`. It holds the panel used to display the background. Content is added to the frame such as `pane.add(MyBackground);`. The `setVisible(true);` method makes sure that the contents are drawn to the screen. This is all wrapped by the curly braces after `public static void`

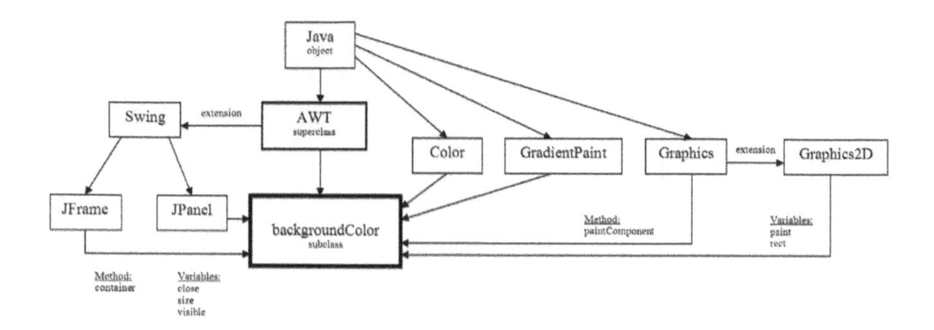

FIGURE 4.3 Inheritance tree graph showing Java at the top of the tree (object), **AWT** (superclass) and **backgroundColor** (subclass), **Swing** and **Graphics2D** (extensions), including inherited methods and variables. © Theodor Wyeld. Used with permission.

`main(String[] args {})`. Without this main routine our class would not do anything.

Now is a good time to talk about inheritance. Inheritance in the context of Java is the notion that the methods of one class can be inherited by another in a hierarchical tree-like structure. At the top of the tree is the super class. For example, the Java base class `awt` is the superclass for our `backgroundColor` class. Our `backgroundColor` class inherits methods from the `awt` class. `Javax.swing` is an extension to the base Java packages, like `awt`. `JFrame` and `JPanel` are subclasses of the `swing` package. We inherit their methods such as `Container` which we are calling 'pane' to paint or draw our background onto: `pane.add(MyBackground);`. We can specify which variables within the inherited frame class we need to access (close, size and visible). `Graphics2D` is an extension of the `Graphics` class. We inherit the method `paintComponent` from the base `Graphics` class while we need to inherit other methods, such as `setPaint` and `fillRect`, from `Graphics2D`. `Color` and `GradientPaint` are subclasses of the `awt` package (see Figure 4.3). Inheritance means you can use the same methods from a superclass over and over without having to actually write the code for the class each time. Whether these are classes that we write ourselves or the standard Java utility classes, this property of the Java programming environment is very useful. One complication to this property, however, is when we write a class that has the same name and properties as a superclass but we want it to behave differently in a particular context. In this instance we need to override the superclasses methods. This occurs in the next class, `TreeFrame`, where a method in the superclass Graphics is being used in a different way.

Now let's add our tree. We will take a slightly different approach this time. Later we will want add some new classes so we need to start organising our classes so they can be accessed from each other. In this case we extend the standard `JFrame` and `JPanel` classes rather than just invoke or create them. The frame is still a container to hold our panel upon which we draw images but we can now use a lot of different methods to generate our graphic content. Having them organised into sub classes also helps us separate out what each of them does.

We have already discussed the instance variables for the parts of our tree that we want to be able to change. These are declared inside the `JPanel Tree` class because this is where they will be used. The `initLeafRandoms()` method sets the random spacing between leaves. It is done only once on the initial `Tree` construction rather than every time the canvas is redrawn to optimise the use of computing resources. The '`void`' in the '`void drawTree`' method just means it does not actually return a value. If it was required to return a value we would use `int`. The `g2d` part inside the brackets is just shorthand for `Graphics2D`. We use `g2d` wherever we refer to the standard `Graphics2D` class. From here we can invoke instances or objects within the `Graphics2D` class. For example, `g2d.setColor(treeColor);` invokes an instance of `setColor` which takes 3 values (for red, green and blue). It uses the values we assigned to `treeColor`. The rest is pretty straightforward. Instances of `Rectangle` are used to create a trunk and branches for our tree which are 'filled' with `treeColor`. In the case of the branches, these are rotated incrementally to form the fan shape of our tree using the standard `Math` class. The actual branches are created by the `drawBranches` method. Similarly, the leaves are created by the `drawLeaves` method. A random number of leaves is drawn around each branch. We set the range of random leaves between 20 and 80 in the `initLeafRandoms` method. And, finally, our main method is invoked after all the calculating and creating is done. The '`new Treeframe`' is a constructor of our main `TreeFrame` class. It is created in the `main()` method.

Listing 4.3 Tree frame

```
1  import java.awt.Color;
2  import java.awt.GradientPaint;
3  import java.awt.Graphics;
4  import java.awt.Graphics2D;
5  import java.awt.Rectangle;
6  import java.awt.geom.Ellipse2D;
7  import java.util.Random;
8  import java.awt.*;
9  import java.awt.event.*;
10 import javax.swing.*;
11 import javax.swing.JPanel;
12
13 public class TreeFrame extends JFrame { // TreeFrame is a subclass of
       JFrame
14
15   private Tree canvas;
16
17   public TreeFrame() {
18     canvas = new Tree();
```

```
19        canvas.setPreferredSize(new Dimension(640, 480));
20
21        Container cp = getContentPane();
22        cp.add(canvas);
23
24        setDefaultCloseOperation(EXIT_ON_CLOSE);
25        pack();
26        setVisible(true);
27      }
28
29      public class Tree extends JPanel { // Tree is a subclass of JPanel
30        int x = 300;
31        int y = 350;
32        int branchingNum = 7;
33        int length = 45;
34        int width = 10;
35        int angle = 25;
36        int leafRandoms[];// a variable that holds an array of type integers
37        Color treeColor = new Color(142, 98, 89);
38        Color leafColor = new Color(62, 142, 73);
39        GradientPaint backgroundColor = new GradientPaint(0, 480, Color.BLACK,
40            0, 240, Color.WHITE);
41
42        public Tree() {
43          initLeafRandoms();
44        }
45
46        @Override
47        public void paintComponent(Graphics g) {
48          super.paintComponent(g);
49          paintBackground((Graphics2D) g);
50          drawTree((Graphics2D) g);
51        }
52
53        private void paintBackground(Graphics2D g) {
54          g.setPaint(backgroundColor);
55          g.fillRect(0, 0, getWidth(), getHeight());
56        }
57
58        void drawTree(Graphics2D g2d) {
59          g2d.setColor(treeColor);
60          Rectangle trunk = new Rectangle(x, y, width, length * 2);
61          Ellipse2D trunkBottom = new Ellipse2D.Double(x, y + (length * 2)
62              - ((width / 2) / 2), width, width / 2);
63          g2d.fill(trunk);
64          g2d.fill(trunkBottom);
65          Rectangle branch = new Rectangle(x, y, width, length);
66          g2d.rotate(Math.toRadians(angle / 2), x, y);
67          drawBranches((Graphics2D) g2d.create(), branch, branchingNum, angle);
68          drawLeaves((Graphics2D) g2d.create(), branch, branchingNum, angle);
69        }
70
71        void drawBranches(Graphics2D g2d, Rectangle baseBranch, int numLeft,
72            double angle) {
73
74          if (numLeft > 0) {
75            numLeft--;
76            Rectangle newBranch = new Rectangle(baseBranch);
77            newBranch.setLocation(newBranch.x, newBranch.y
78                - newBranch.height);
79            g2d.rotate(Math.toRadians(-angle), baseBranch.getX(),
80                baseBranch.getY());
81            g2d.fill(newBranch);
82            drawBranches(g2d, newBranch, numLeft, angle);
83            g2d.rotate(Math.toRadians(angle), baseBranch.getX(),
84                baseBranch.getY());
85            g2d.fill(newBranch);
```

```
86            drawBranches(g2d, newBranch, numLeft, -angle);
87          }
88        }
89
90        void drawLeaves(Graphics2D g2d, Rectangle baseBranch, int numLeft,
91            double angle) {
92
93          g2d.setColor(leafColor);
94          if (numLeft > 0) {
95            numLeft--;
96            Rectangle newBranch = new Rectangle(baseBranch);
97            newBranch.setLocation(newBranch.x, newBranch.y
98                - newBranch.height);
99            g2d.rotate(Math.toRadians(-angle), baseBranch.getX(),
100               baseBranch.getY());
101           drawLeaves(g2d, newBranch, numLeft, angle);
102           g2d.rotate(Math.toRadians(angle), baseBranch.getX(),
103               baseBranch.getY());
104           drawLeaves(g2d, newBranch, numLeft, -angle);
105         }
106         if ((numLeft + 0.0) / branchingNum < 0.7) {
107           Random random = new Random();
108           for (int i = 0; i < 6; i++) { // i increments up to 6 times
109             g2d.fillOval((int) baseBranch.getX() - 40 + leafRandoms[i]
110                 + (random.nextInt(8) - 4),
111                 (int) baseBranch.getY() - 40 + leafRandoms[i + 5]
112                     + (random.nextInt(8) - 4), 10, 20);
113           }
114         }
115       }
116
117       void initLeafRandoms() {
118         Random random = new Random();
119         leafRandoms = new int[20]; // new array of leafRandoms with 20
120                       // elements
121         for (int i = 0; i < leafRandoms.length; i++) {
122           leafRandoms[i] = random.nextInt(80);
123         }
124       }
125     }
126
127     public static void main(String[] args) {
128
129       SwingUtilities.invokeLater(new Runnable() {
130         public void run() {
131           new TreeFrame();
132         }
133       });
134     }
135   }
```

To generate the leaves on the tree in our `TreeFrame` class we are using an array. An array is a way to store a list or collection of elements, all of the same type (integers, text, other arrays, and objects). In this case, when we declare the variable `leafRandoms` we do so as an array. The `int` art refers to the type and the square brackets indicate that this is an array – meaning it will have multiple values. We are using the array to draw random leaves about the length of our branches. This is initiated at the `initLeafRandoms()` method. First we need to create a new array object: `leafRandoms = new int [20];` and assign it to `leafRandoms` with 20 elements. Next we use a For loop to repeat a statement a number of times until a specified condition is matched. We start the loop by initialising its index `i`. Starting as 0, we need to test if we have matched the end condition yet. To do this we compare `i` with `leafRandoms`.length to see that it is less than the specified number of array elements (20). If not true, the statement inside the curly braces {} executes another loop. Once the test is false it stops. This is incremented using the `i++` expression. In this case, the statement is 'for each iteration of `leafRandoms` assign a random number between 0 and 80'. In real terms this means draw 20 + 5 leaves along each length of a branch at random locations and orientations. Why 25 and not 20? Under the `drawLeaves()` method you will notice that `leafRandoms[i+5]` is called. The +5 adds 5 to the 20 leaves generated at the `initleafRandoms()` method.

Unlike the previous basic class `backgroundColor`, in the `TreeFrame` class some arithmetic was used to generate particular values for generating the number of branches and leaves. In Java, these are called expressions and operators. Expressions are statements that return a value. Operators are the symbols used in an expression as part of the arithmetic equation (+ - * / %, note: % is modulus not percent). For example, where the `trunkBottom` is created we have used `Ellipse2D.Double()` to specify its parameters. The ellipse class creates rectangles – using the '`Double`' just means we have more precise coordinates than if we just used `Ellipse2D` on its own (which uses whole integers only). The parameters for `Ellipse2D.Double()` include the `x` and `y` coordinates for the upper-left corner of our rectangle, and the `width` and `height` of our rectangle. By using the expression `y+(length*2)-((width/2)/2)` for the y coordinate of the upper-left corner of our rectangle we can change the length and width variables later and it will affect where the y coordinate results. Similarly, using the expression `width/2` for the height of our rectangle we know that the proportion of width to height will be constant when we change the width variable. Another type of expression that returns a value is the use of 'dot notation'. For example, `g2d.setColor(treeColor);` returns the colour `treeColor` to `g2d`. In this case, it colours the tree's trunk `treeColor` (brown). If you save the above code in a text file called `TreeFrame.java`, compile and run it, you should see the screen in Figure 4.4.

4.3.2 Adding some controls

We could put all the methods for our class and subclasses in a single Java file, but it is always good practice to separate out the core functions into their own class files. Therefore, we will now create three separate files: `Canvas`, `Tree` and `ControlFrame`. We will use `Canvas` as our main class which references our tree class and control frame class. The control frame class will call up all the controls we need to manipulate the various components of our tree class. The canvas class connects the tree class variables by specifying what controls will be used (sliders, buttons and so on) while the control frame class specifies how they will be laid out and their default parameters. The variables we have

FIGURE 4.4 Output of TreeFrame class. © Theodor Wyeld. Used with permission.

access to in the tree class are: Position (x and y), branching number, branch overall length, width and angle, and branch length and width reduction. It is in the canvas class that we use the main routine to launch our program. But, instead of wrapping all our sub classes in a single instance of main() we can place it on its own as a method so it can be accessed by outside classes.

```
1  public static void main(String[] args) {
2    javax.swing.SwingUtilities.invokeLater(new Runnable() {
3      public void run() {
4        createAndShowGUI();
5      }
6    });
7  }
```

Because our program is a graphics program we need to create a window to display our graphics objects in. Both the Java AWT and Swing are windowing toolkits. However, they are quite different in how they can be used as graphics programs. Without going into too much explanation the AWT (abstract windowing kit) is limited in what it can do graphically. It is mostly used for creating standard graphics user interface applications such as Microsoft Word (although clearly MS Word is not a Java application!). MS Word is used as an example here because it does not typically involve any heavy graphics like image manipulation. Nonetheless, all the buttons, icons, and text generating and editing involve some graphics processing. But, compared to what you can do in Adobe Photoshop, MS Word is not a refined graphics package. On the other hand, Swing allows us finer control over how graphics are handled. It allows for manipulation of a graphics image at the pixel level. Although some of this functionality is available in the AWT API, Swing is more fully-featured.

That is why we are using it here. Both `AWT` and `Swing` support four main types of objects: Containers (such as panel), canvases (although we can draw on a panel canvases are better at handling images), user-interface components (labels, buttons, textfields, sliders and so on), and window construction components (windows, frames, menus, dialogs and so on). At the root of most `AWT` and `Swing` functionality is the Components class.

A core feature of all graphics programs is how images are painted or drawn to the screen. The order in which images are drawn is important – the ordering of what should be drawn first, in front of or on top of another image. To do this it is important to execute the drawing functions in the correct sequence. Swing does this by using only a single thread. A thread is where the code logic unfolds. This is how it tells the machine what tasks to do and in which order to do them. Typically it is the single thread that invokes the main method of the program class. In `Swing` programs, the initial thread creates a `Runnable` object that initialises (or starts) the GUI and schedules it for execution on the event dispatch thread (EDT). Once the GUI is created, the program can be driven by GUI events. Each event causes the execution of a short task on the EDT. Additional tasks can be scheduled on the event dispatch thread and released between the main event processing. This allows the main window to be constantly updated in an orderly manner.

Our initial thread is started with '`class Canvas extends JPanel`' and ended or destroyed with:

```
frame.setDefaultCloseOperation(JFrame.EXIT_ON_CLOSE);
```

By placing the `main()` method at the end of the class statement this gives the program time to build the GUI in the correct order before executing it. By including the `invokeLater` call, the run method can schedule events, such as from buttons, sliders, textfields and so on. The buttons and sliders are called up in the `Canvas` class but their parameters or variables are defined in our `ControlFrame` class.

While the `ControlFrame` class may only use a single thread many applications use multiple threads. Using multiple threads means different processes can be running at the same time instead of having to wait for one process to end before the next can start. This means the results of other processes can be available at the same time. This is particularly pertinent with animation loops. For example, to build a simple clock application in Java we need a '`runnable`' interface to run a thread. This is achieved using the `run()` method. Next we need an instance variable of type Thread which we will call runner to hold the program's thread. Then we need to start our thread using the `start()` method. Finally, we need to add a `stop()` method to terminate the execution when the program is closed – otherwise it will continue to run. The `stop()` method not only stops the thread from executing, but it also sets the thread's variable (`runner`) to `null`. This means the Thread object can be removed from memory. Figure 4.5 shows the output.

Listing 4.4 Clock

```
import java.awt.*;
import javax.swing.*;
import java.util.*;

class Clock extends JFrame implements Runnable {
  Thread runner; // global object - instance variable
  Font clockFont;

  public Clock() {
    super("Java clock");
    setSize(350, 100);
    setDefaultCloseOperation(JFrame.EXIT_ON_CLOSE);
    setVisible(true);
    setResizable(false); // create window

    clockFont = new Font("Serif", Font.BOLD, 40);

    Container contentArea = getContentPane();
    ClockPanel timeDisplay = new ClockPanel();

    contentArea.add(timeDisplay);
    setContentPane(contentArea);
    start(); // start thread running

  }

  class ClockPanel extends JPanel {
    public void paintComponent(Graphics painter) {

      painter.setFont(clockFont);
      painter.setColor(Color.black);
      painter.drawString(timeNow(), 60, 40);
    }
  }

  public String timeNow()// get current time
  {
    Calendar now = Calendar.getInstance();
    int hrs = now.get(Calendar.HOUR_OF_DAY);
    int min = now.get(Calendar.MINUTE);
    int sec = now.get(Calendar.SECOND);

    String time = hrs + ":" + min + ":" + sec;

    return time;
  }

  public void start() {
    if (runner == null)
      runner = new Thread(this);
    runner.start();// method to start thread
  }

  public void run() {
    while (runner == Thread.currentThread()) {
      repaint();

      try // define thread task
      {
        Thread.sleep(1000);
      } catch (InterruptedException e) {
        System.out.println("Thread failed");
      }
    }
  }

```

FIGURE 4.5 Clock. © Theodor Wyeld. Used with permission.

```
67    public void stop() {
68      if (runner != null) {
69        runner.stop();
70        runner = null;
71      }
72    }
73
74    public static void main(String[] args) // create main method
75    {
76      Clock clock = new Clock();
77    }
78  }
```

There are many different methods for laying out the graphics components of a GUI program (displaying the tree). The four most commonly used are: Grid, border, bow and flow. They each handle the graphics components in different ways. They can be combined and embedded in one another. Our `Canvas` class uses the `setLayout` method to define `BoxLayout`. For the ease of laying out our components the `BoxLayout` is used for both the Canvas class and `ControlFrame` class. `BoxLayout` simply stacks components on top of each other in a column or next to each other in a row. They can be combined to create both columns and rows. We specify how `BoxLayout` handles the graphics components by using calls such as `BoxLayout.PAGE_AXIS` (which lays out the components as you would words on a page: From left to right and top to bottom). `BoxLayout.Y_AXIS` is used in the `ControlFrame` class to align all the controls in a list from top to bottom of the panel. Our `layeredPane` refers to the way the various components of our tree are drawn. As the name suggests, each component is laid on top of the previous. In the `Tree` class we see that the background is drawn first, then trunk, branches and finally the leaves. If they were in any other order some components would be hidden by others. Our `ControlFrame` is launched in a separate window. After the main GUI is launched the `ControlFrame` is added. The canvas class calls up the controls that will populate the `ControlFrame` panel. Each control called up in `Canvas` includes the various variables available for manipulation. For example, to change the number of branches in our tree we use the `Branching Number` control. This is specified in the `Canvas` class using the below method:

```
1  controlFrame.addSlideControl("Branching Number", 0, 10, tree.branchingNum,
       1, a -> {
2    tree.branchingNum = ((JSlider)a.getSource()).getValue();
3    tree.repaint();
4  });
```

The method includes access to its location (**controlFrame**), the type of control (**addSlideControl**), what attribute of the tree is being manipulated by the control ("**Branching Number**"), and the range of change (0, 10). The **->** is a lambda expression used to separate the parameters on the left-side from the actual expression inside the curly braces on the right side. In this case, from **tree.branchingNum**, **a.getSource()** is a member of the **branchingNum**. It returns a value between 0 and 10. The **tree.repaint()** tells the panel to update and redraw the tree with the new value after the control has been adjusted (when the slider is moved by a mouse cursor) and returned a value. The final "Default Values" button invokes the **tree.resetTreeValues** method returning the tree display to its default parameters which it finds in the **Tree** class.

Listing 4.5 Canvas

```
1  import java.awt.Dimension;
2
3  import javax.swing.BoxLayout;
4  import javax.swing.JComponent;
5  import javax.swing.JFrame;
6  import javax.swing.JLayeredPane;
7  import javax.swing.JPanel;
8  import javax.swing.JSlider;
9  import javax.swing.JToggleButton;
10
11 class Canvas extends JPanel { // Canvas is a subclass of JPanel
12   private JLayeredPane layeredPane;
13
14   public Canvas() {
15     setLayout(new BoxLayout(this, BoxLayout.PAGE_AXIS));
16     layeredPane = new JLayeredPane();
17     layeredPane.setPreferredSize(new Dimension(600, 600));
18
19     Tree tree = new Tree();
20
21     tree.setBounds(0, 0, 600, 600);
22     layeredPane.add(tree);
23     add(layeredPane);
24     ControlFrame controlFrame = new ControlFrame();
25     initControlComponents(controlFrame, tree);
26     controlFrame.setBounds(1250, 50, 300, 700);
27     controlFrame.setVisible(true);
28   }
29
30   public static void createAndShowGUI() {
31     JFrame frame = new JFrame();
32     frame.setDefaultCloseOperation(JFrame.EXIT_ON_CLOSE);
33     frame.setLocation(600, 50);
34     frame.setResizable(false);
35
36     JComponent newContentPane = new Canvas();
37     newContentPane.setOpaque(true);
38     frame.setContentPane(newContentPane);
39
40     frame.pack();
41     frame.setVisible(true);
42   }
```

```
43
44    public static void main(String[] args) {
45      javax.swing.SwingUtilities.invokeLater(new Runnable() {
46        public void run() {
47          createAndShowGUI();
48        }
49      });
50    }
51
52    public static void initControlComponents(ControlFrame controlFrame, Tree
          tree) {
53
54      controlFrame.addHeading("Tree Controls");
55
56      controlFrame.addSlideControl("Width", 0, 100, tree.width, a -> {
57        tree.width = ((JSlider)a.getSource()).getValue();
58        tree.repaint();
59      });
60
61      controlFrame.addSlideControl ("Length", 0, 200, tree.length, a -> {
62        tree.length = ((JSlider)a.getSource()).getValue();
63        tree.repaint();
64      });
65      controlFrame.addSlideControl("X", 0, 600, tree.x, a -> {
66        tree.x = ((JSlider)a.getSource()).getValue();
67        tree.repaint();
68      });
69      controlFrame.addSlideControl("Y", 0, 600, tree.y, a -> {
70        tree.y = ((JSlider)a.getSource()).getValue();
71        tree.repaint();
72      });
73      controlFrame.addSlideControl("Length Reduction", 0, 10,
            tree.lReduction, 1, a -> {
74        tree.lReduction = ((JSlider)a.getSource()).getValue();
75        tree.repaint();
76      });
77      controlFrame.addSlideControl("Width Reduction", 0, 10,
            tree.wReduction, 1, a -> {
78        tree.wReduction = ((JSlider)a.getSource()).getValue();
79        tree.repaint();
80      });
81      controlFrame.addSlideControl("Branching Number", 0, 10,
            tree.branchingNum, 1, a -> {
82        tree.branchingNum = ((JSlider)a.getSource()).getValue();
83        tree.repaint();
84      });
85      controlFrame.addSlideControl("Angle", 0, 180, tree.angle, 45, a -> {
86        tree.angle = ((JSlider)a.getSource()).getValue();
87        tree.repaint();
88      });
89
90      controlFrame.addButtonControl("Default Values", a ->
            tree.resetTreeValues());
91
92
93
94    }
95  }
```

The `ControlFrame` defines the parameters for the controls we will use to manipulate our tree. The controls are displayed on a framed panel in its own window separate from the main Canvas panel which displays our tree. The panel is scrollable. Only three types of graphic components are specified (text, slider and button) of which only two are controllers (slider and button). The text parameters are handled by the **addHeading** method, slider by **addSlideControl**, and button by **addButtonControl**. Specifying these controls only once means the different controls on the **ControlFrame** panel do not have to be specified for each control; the same method can be reused by the different controls. The **addSlideControl** method is referred to twice because in the first instance the label, range, and a change listener is added. In the second instance with these in place, the default parameters are specified. This means that different types of sliders can be generated without having to write all the features of a slider that might not be needed (**tickSpacing** and so on). Figure 4.6 shows the output.

Listing 4.6 Control frame

```
1   import java.awt.Color;
2   import java.awt.Font;
3   import java.awt.event.ActionListener;
4
5   import javax.swing.Box;
6   import javax.swing.BoxLayout;
7   import javax.swing.JButton;
8   import javax.swing.JCheckBox;
9   import javax.swing.JFrame;
10  import javax.swing.JLabel;
11  import javax.swing.JPanel;
12  import javax.swing.JScrollPane;
13  import javax.swing.JSeparator;
14  import javax.swing.JSlider;
15  import javax.swing.JTextField;
16  import javax.swing.event.ChangeListener;
17
18  class ControlFrame extends JFrame { // ControlFrame is a subclass of JFrame
19
20    private JPanel container;
21    private Font headingFont = new Font("Sans-Serif", Font.BOLD, 16);
22
23    public ControlFrame() {
24      setTitle("Controls");
25      setDefaultCloseOperation(JFrame.EXIT_ON_CLOSE);
26      container = new JPanel();
27      container.setLayout(new BoxLayout(container, BoxLayout.Y_AXIS));
28      this.add(new JScrollPane(container));
29    }
30
31    public void addSeperator() {
32      container.add(new JSeparator());
33    }
34
35    public void addHeading(String text) {
36      JLabel heading = new JLabel(text);
37      heading.setFont(headingFont);
38      heading.setAlignmentX(LEFT_ALIGNMENT);
39      container.add(heading);
40      container.add(Box.createVerticalStrut(15));
41      pack();
42    }
43
44    public void addSlideControl(String label, int min, int max, int value,
45        ChangeListener cl) {
46      addSlideControl(label, min, max, value, 100, cl);
47    }
48
49    public void addSlideControl(String text, int min, int max, int value,
```

```
50          int tickSpacing, ChangeListener a) {
51        JSlider slider = new JSlider(min, max, value);
52        slider.setMajorTickSpacing(tickSpacing);
53        slider.setPaintTicks(true);
54        slider.setPaintLabels(true);
55        slider.addChangeListener(a);
56        slider.setBackground(new Color(225, 225, 250));
57        slider.setAlignmentX(LEFT_ALIGNMENT);
58        JLabel label = new JLabel(text);
59        label.setAlignmentX(LEFT_ALIGNMENT);
60        container.add(label);
61        container.add(slider);
62        container.add(Box.createVerticalStrut(15));
63        pack();
64      }
65
66      public void addButtonControl(String label, ActionListener al) {
67        JButton button = new JButton(label);
68        button.addActionListener(al);
69        button.setAlignmentX(LEFT_ALIGNMENT);
70        container.add(button);
71        container.add(Box.createVerticalStrut(15));
72        pack();
73      }
74    }
```

4.3.3 Controlling Tree class with ControlFrame from Canvas class

To manipulate the elements of our Tree class from the control panel we need to integrate the Tree class with the Canvas class. To do this we need to change our earlier `TreeFrame` class from a standalone app to a class which is launched by the Canvas class. Where `TreeFrame` class created its own window or frame within which the panel was generated for displaying our tree now the frame is created by the `Canvas` class. Therefore our new Tree class only needs to specify a panel – class `Tree` extends `JPanel`. Apart from this much of the code remains largely unchanged. At the end of the code, the run argument is removed (this now occurs from the Canvas class instead) and replaced with some default values we can use to reset our controls.

Listing 4.7 Tree

```
1   import java.awt.Color;
2   import java.awt.GradientPaint;
3   import java.awt.Graphics;
4   import java.awt.Graphics2D;
5   import java.awt.Rectangle;
6   import java.awt.geom.Ellipse2D;
7   import java.util.Random;
8
9   import javax.swing.JPanel;
10
11  class Tree extends JPanel { // Tree is a subclass of JPanel
12    int x = 300;
13    int y = 350;
14    int branchingNum = 7;
15    int length = 85;
16    int width = 25;
17    int angle = 30;
18    int lReduction = 4;
```

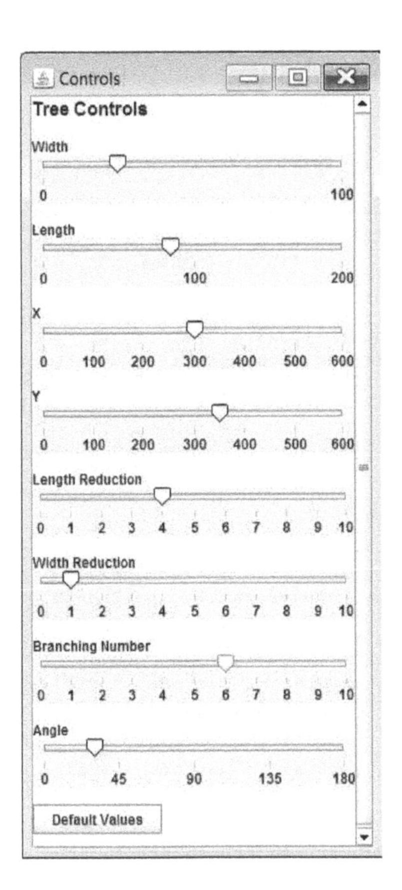

FIGURE 4.6 ControlFrame. © Theodor Wyeld. Used with permission.

```
19      int wReduction = 1;
20      int leafRandoms[];
21      Color treeColor = new Color(142, 98, 89);
22      Color leafColor = new Color(62, 142, 73);
23
24      public Tree() {
25        initLeafRandoms();
26      }
27
28      public void paintComponent(Graphics g) {
29        GradientPaint backgroundColor = new GradientPaint(0, 480, Color.BLACK,
30            0, 240, Color.WHITE);
31        Graphics2D g2D = (Graphics2D) g;
32        g2D.setPaint(backgroundColor);
33        g2D.fillRect(0, 0, getWidth(), getHeight());
34        drawTree((Graphics2D) g);
35      }
36
37      void drawTree(Graphics2D g2d) {
38        g2d.setColor(treeColor);
39        Rectangle trunk = new Rectangle(x, y, width, length * 2);
40        Ellipse2D trunkBottom = new Ellipse2D.Double(x, y + (length * 2)
41            - ((width / 2) / 2), width, width / 2);
42        g2d.fill(trunk);
43        g2d.fill(trunkBottom);
44        Rectangle branch = new Rectangle(x, y, width, length);
45        g2d.rotate(Math.toRadians(angle / 2), x, y);
46        drawBranches((Graphics2D) g2d.create(), branch, branchingNum, angle);
47        drawLeaves((Graphics2D) g2d.create(), branch, branchingNum, angle);
48      }
49
50      void drawBranches(Graphics2D g2d, Rectangle baseBranch, int numLeft,
51          double angle) {
52        // draw branches
53        if (numLeft > 0) {
54          numLeft--;
55          Rectangle newBranch = new Rectangle(baseBranch);
56          newBranch.grow(-wReduction, -lReduction);
57          newBranch.setLocation(newBranch.x, newBranch.y - newBranch.height);
58          g2d.rotate(Math.toRadians(-angle), baseBranch.getX(),
59              baseBranch.getY());
60          g2d.fill(newBranch);
61          drawBranches(g2d, newBranch, numLeft, angle);
62          g2d.rotate(Math.toRadians(angle), baseBranch.getX(),
63              baseBranch.getY());
64          g2d.fill(newBranch);
65          drawBranches(g2d, newBranch, numLeft, -angle);
66        }
67      }
68
69      void drawLeaves(Graphics2D g2d, Rectangle baseBranch, int numLeft,
70          double angle) {
71        // draw leaves
72        g2d.setColor(leafColor);
73        if (numLeft > 0) {
74          numLeft--;
75          Rectangle newBranch = new Rectangle(baseBranch);
76          newBranch.grow(-wReduction, -lReduction);
77          newBranch.setLocation(newBranch.x, newBranch.y - newBranch.height);
78          g2d.rotate(Math.toRadians(-angle), baseBranch.getX(),
79              baseBranch.getY());
80          drawLeaves(g2d, newBranch, numLeft, angle);
81          g2d.rotate(Math.toRadians(angle), baseBranch.getX(),
82              baseBranch.getY());
83          drawLeaves(g2d, newBranch, numLeft, -angle);
84        }
85        if ((numLeft + 0.0) / branchingNum < 0.7) {
```

```
86          Random random = new Random();
87          for (int i = 0; i < 10; i++) {
88            g2d.fillOval((int) baseBranch.getX() - 40 + leafRandoms[i]
89                + (random.nextInt(8) - 4), (int) baseBranch.getY() - 40
90                + leafRandoms[i + 5] + (random.nextInt(8) - 4), 10, 20);
91          }
92        }
93      }
94
95      void initLeafRandoms() {
96        Random random = new Random();
97        leafRandoms = new int[20];
98        for (int i = 0; i < leafRandoms.length; i++) {
99          leafRandoms[i] = random.nextInt(80);
100       }
101     }
102
103     void resetTreeValues() {
104       x = 300;
105       y = 350;
106       branchingNum = 7;
107       length = 85;
108       width = 25;
109       angle = 30;
110       lReduction = 4;
111       wReduction = 1;
112     }
113   }
```

If you now compile and execute all three class files (`Canvas`, `Tree` and `ControlFrame`) you should be able to control your tree using the control panel.

4.3.4 Connecting to a live feed

Now that we have a tree which has elements that can be manipulated by controls (sliders) we could also use other types of input to affect the tree's elements. As discussed earlier there are many different types of live feed data that are streamed across the internet. The data takes many different forms and values. We can use some of these to influence how the elements of our tree behave. A common live feed is what comes from a weather station. Among other types of data, a weather station feed includes temperature, wind speed and direction, rain and so on. All of these can be used to affect how our tree behaves. It just takes a little imagination to work out what type of data should affect what part of our tree. In the following example, we are going to use temperature to affect our background colouring; wind speed and direction to affect our tree and its branches; and we are going to produce some rain. There are many more effects we could include, but these three are sufficient to demonstrate the possibilities. But first we need to be able to get the data from the weather station in a form that we can use.

While there are many live weather data feed sites, `openweather.com` provides an easy to implement API (Application Programming Interface). By connecting to `openweather.com`'s API we can get live feed data in a form that is useful. To do this we need to create an `OpenWeather` class to handle

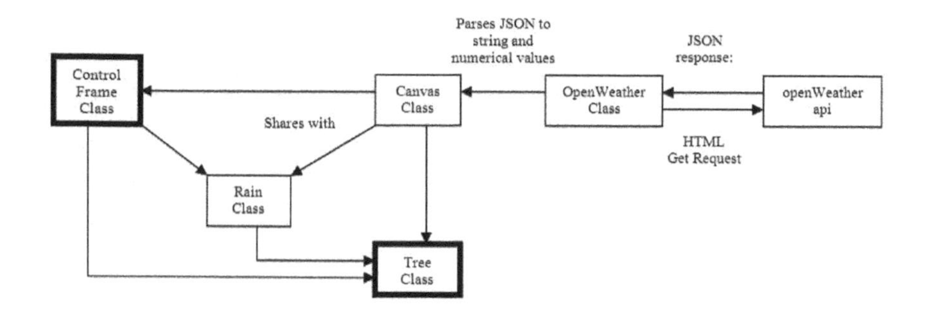

FIGURE 4.7 Diagram outlining how the overall application works. © Theodor Wyeld. Used with permission.

the communications and parse (from packet data to strings of numbers or text) the data onto our canvas class in a form that it understands (text and numbers). These are then organised such that they affect those elements of our tree object otherwise controlled by our sliders. We are effectively substituting the output from our sliders with input from the `OpenWeather` class file. A diagram outlining how the overall application works is shown below (see Figure 4.7).

4.3.5 Connecting to OpenWeather.com

To connect to the `openweather.com` API we need to create a class file to establish a connection, communicate with the remote site across the net, and parse the information to `Canvas` in a form that it understands and can use. To do this we need to include some standard Java classes from the `io` and `net` packages and `JSON` (JavaScript Object Notation). JSON is a data-interchange format. Based on JavaScript it converts dense data into human-readable information. The various `io`, `net` and `JSON` classes connect to the openweather server, retrieve data, store it temporarily, sort it into arrays and objects, and finally parse it to the `Canvas` class. The `OpenWeather` class defines what information is to be retrieved from `openweather.com`. It includes location, temperature, `WindSpeed`, `WindDirection`, and `Rain`[1].

Using the standard `io` and `net` Java classes along with `JSON`, the `OpenWeather` class constructor makes a request to the openweather API for `name`, `temp`, `windspeed`, `winddirection` and if it is raining (`Boolean` state). The first method invokes our `OpenWeather` class and makes the request. The

[1]To run this properly, you will need your own API key from `openweathermap.org` A key is provided in the code example, but keys are updated periodically so it probably will not work any more. Just substitute your key for the one given by inserting between `&APPID=` and `&units=metric`.

second method is the `requestWeather` itself which provides a location for the API to return weather stats for. Then a thread is created to the `openweather` URL which includes our license key. This is where `JSON` acts more like a Python dictionary than an SQL query. When `OpenWeather()` makes the `HTTP GET` request it passes in three parameters: The city name, APPID (for authentication) and a units flag asking it send back information in metric units. The data received is a complete weather summary of that city. By parsing it into a `JSONObject()` we can choose what information we want from it. Because we cannot be sure what type of data will be returned (text, integer, fractions), all information is parsed as a `String` first and then the result is parsed as a `double`. This avoids errors in type. It requests data in the form of strings for known variables listed in the API. These are country, rain, temp, wind speed, and wind deg. Next, we put in a `catch` in case there is an error in our request or on the API side. If there is an error it gets printed to the console. Next, we take our returned values for rain, temp, wind etc. and convert them into strings and numbers. Finally, we can print all the information to the console with the `String toString` method. Although we are not using the `println()` method for our final console display we are combining 'literal' text inside double quotes with returned values as text. For example, `"\nLocation:" + name`, means 'add the value retuned by name to the word Location'. The plus symbols are used to concatenate the string elements into a single output. The `\n` part is called an escape code. It is a non-printable character combination which tells the program to create a new line of text (like a carriage return). Another type of concatenation is where we add vales on one side of an equation to the initial value. For example, the expression `x += y` is shorthand for `x = x + y`. In this example, `x` is added to `y` and the result is assigned to whatever is on the left of the = symbol (`x`). In our `openWeather` class where we find `weather += (...);` followed by some other values in brackets, this means 'add weather to whatever is returned by whatever is in the brackets' to weather and then assign the final result to `weather`'.

Listing 4.8 OpenWeather

```
1  import java.io.BufferedReader;
2  import java.io.FileNotFoundException;
3  import java.io.InputStreamReader;
4  import java.net.ConnectException;
5  import java.net.URL;
6
7  import org.json.simple.JSONArray;
8  import org.json.simple.JSONObject;
9  import org.json.simple.parser.JSONParser;
10
11 public class OpenWeather {
12   // Default values in the instance of network error
13   private String name = "Morphett Vale";
14   private double temperature = 30;
15   private double windSpeed = 15;
16   private double windDirection;
17   private boolean rain;
18   private String weather = "clear";
19
```

```
20      public OpenWeather() {
21        requestWeather();
22      }
23
24      public void requestWeather() {
25        requestWeather(name);
26      }
27
28      public void requestWeather(String name){
29        JSONParser parser = new JSONParser();
30        try {
31          URL url = new URL("http://api.openweathermap.org/data/2.5/weather?q=
32    " + name + "&APPID=InsertYourKeyHere&units=metric");
33          BufferedReader br = new BufferedReader(new
                  InputStreamReader(url.openStream()));
34          JSONObject response = (JSONObject) parser.parse(br.readLine());
35          this.name = (String)response.get("name") + ",
36    " + (String)((JSONObject)response.get("sys")).get("country");
37          rain = false;
38          weather = "";
39          for(Object weatherTypes : (JSONArray)response.get("weather")){
40          weather +=
                  ((JSONObject)weatherTypes).get("main").toString().toLowerCase()
                  + " ";
41            if(weather.contains("rain"))
42              rain = true;
43          }
44    temperature =
45    Double.parseDouble(((JSONObject)response.get("main")).get("temp").toString());
46    windSpeed =
47    Double.parseDouble(((JSONObject)response.get("wind")).get("speed").toString())
          * 1.61;
48      windDirection =
49    Double.parseDouble(((JSONObject)response.get("wind")).get("deg").toString());
50
51        } catch (Exception ex) {
52          if(ex instanceof ConnectException)
53    System.out.println("Network Error: Using default weather object...");
54          else if (ex instanceof FileNotFoundException)
55    System.out.println("City Not Found Error: Using default weather
         object...");
56          else ex.printStackTrace();
57        }
58        System.out.println(this);
59      }
60
61      public String getName() {
62        return name;
63      }
64
65      public double getTemperature() {
66        return temperature;
67      }
68
69      public double getWindSpeed() {
70        return windSpeed;
71      }
72
73      public double getWindDirection() {
74        return windDirection;
75      }
76
77      public boolean isRaining() {
78        return rain;
79      }
80
```

```
81    @Override // used to hide superclass' method where the subclass has same
          signature
82    // (name, number and type of arguments) as a method in the superclass
83    public String toString(){
84      return "___________________________________\n" +
85          "Weather"
86          + "\nLocation: " + name
87          + "\nTemp: " + temperature
88          + "\nWeather: " + weather
89          + "\nWindDirection:" + windDirection
90          +"\n___________________________________";
91    }
92  }
```

Note: To get this to compile and run, you need to open DOS or the command prompt window, navigate to the folder that holds your `weathertree` Java files, make sure a copy of `json-simple-1.1.1.jar` is in the Java folder[2]. Then, to compile use the following command:

```
javac -classpath json-simple-1.1.1.jar *.java
```

If all your files compile without errors, then to run use the following command < insert the folder address that contains the your class files between parentheses >:

```
java -classpath
    "<thisFolder'sPath>;<thisFolder'sPath>\json-simple-1.1.1.jar" Canvas
```

for example:

```
java -classpath "C:\myFolder\; C:\myFolder\json-simple-1.1.1.jar" Canvas
```

or, here is an example of how to run the command in Linux:

```
java -classpath "/home/myFolder; /home/myFolder/json-simple-1.1.1.jar"
    Canvas
```

4.3.6 Accessing the data from Canvas class

In order to use the information provided by the `openweather` API and parsed by our `JSON OpenWeather` class we need to add some code to our `Canvas` class. The first thing we need to do is include a timer so we can update our information in real time. To do this we need to import `javax.swing.Timer;`. This gets called up in our main class: `Canvas` extends `JPanel`. `Timer timer;`

[2]It can be downloaded from here:
`http://www.java2s.com/Code/Jar/j/Downloadjsonsimple111jar.htm`

is made private because we only need to access it from within this class file execution.

Next we want to add all the featured data we have requested from the **openweather** API – **rain**, **temp**, **wind**. To do this, inside our main **Canvas** method we include a reference to **Rain rain = new Rain(600);**. Still inside the **Canvas** method, we need to invoke our **OpenWeather** class to access the information it parses and pass it to the elements of our tree we that we want the information to affect. For example, we are passing the temperature values to the sky colour, wind direction to how much the tree sways and in which direction, and whether it is raining or not. Finally, we need to refresh our data stream so that it is always current and passed to the elements of our tree in real-time.

```
1   OpenWeather weather = new OpenWeather();
2   Timer weatherRefresh = new Timer(600000, a -> { //every 10 minutes
3     weather.requestWeather();
4     tree.calculateSkyColor(weather.getTemperature());
5     tree.calculateSwayBias(weather.getWindDirection());
6     rain.setWind(weather.getWindDirection());
7     rain.setVisible(weather.isRaining());
8   });
9
10  tree.calculateSkyColor(weather.getTemperature());
11  tree.calculateSwayBias(weather.getWindDirection());
12  rain.setWind(weather.getWindDirection());
13  rain.setVisible(weather.isRaining());
14
15  weatherRefresh.start();
```

It is inside the main **Canvas** method that the tree is specified. Therefore we need to specify also our new class Rain. This is specified as part of a layered pane so the rain is drawn on top of the tree, even though the rain is a class outside the tree class. The bounds refer to canvas size. Finally, we specify how often our layered pane is updated: 100.

```
1   tree.setBounds(0, 0, 600, 600);
2   rain.setOpaque(false);
3   rain.setBounds(0, 0, 600, 600);
4   layeredPane.add(rain);
5   layeredPane.add(tree);
6   add(layeredPane);
7
8   timer = new Timer(100, a -> layeredPane.repaint());
9   timer.start();
```

At the end of our main Canvas method we initialise our control frame class and add the control frame itself, tree, weather data and rain graphics.

```
1   //Control Frame for override controls
2   ControlFrame controlFrame= new ControlFrame();
3   initControlComponents(controlFrame, tree, weather, rain);
```

It is in the main **Canvas** class that we define our controls (sliders and so on) and their parameters (range and what they affect). Therefore, we can add controls for our **openweather** data here also. Their initial values come from the weather data feed, but we can override this with our controls to simulate how other values might affect the elements of our tree. We need to include the **OpenWeather** and **Rain** classes inside the arguments for our **initControlComponents**. Then the first control we add is a textfield for typing the city we want the **openweather** API to provide weather data for. This data is printed to the console to check that we have a legitimate feed. Next we add a slider for wind direction which is passed to the sway bias method in the **Tree** class. **Repaint** is added so that after changes are made by either the **openweather** data or from our slider the tree is updated. A similar method is used to add rain. This is a **Boolean** method – either on or off. If the weather data feed shows no rain then we can manually set it to rain using a button. We can then use a slider to set the amount of rain. Similarly, the 24hr range of temperatures for the city selected are used to generate a colour ramp for our background. A slider is defined so we can simulate other temperature ranges.

Listing 4.9 Initialise control components

```
public static void initControlComponents(ControlFrame controlFrame, Tree
    tree, OpenWeather weather, Rain rain){

  controlFrame.addHeading("City Search");
      controlFrame.addSubmissionBox("Set City", weather, tree, rain);
      controlFrame.addButtonControl("Print Weather To Console",
      a -> System.out.println(weather));

      controlFrame.addSeperator();

      controlFrame.addHeading("Weather Controls");
      controlFrame.addSlideControl("Wind Direction", 0, 360,
          (int)weather.getWindDirection(), 45,
      a -> {
        rain.setWind(((JSlider)a.getSource()).getValue());
        tree.calculateSwayBias(((JSlider)a.getSource()).getValue());
        tree.repaint();
      });
      controlFrame.addCheckBoxControl("Rain Toggle", weather.isRaining(),
          a -> {
        rain.setVisible(((JToggleButton)a.getSource()).isSelected());
        tree.repaint();
      });

      controlFrame.addSlideControl("Rain Amount", 0, 1000, 600, 200, a -> {
        rain.setDrops(((JSlider)a.getSource()).getValue());
        tree.repaint();
      });

      controlFrame.addSlideControl("Temperature (C)", 0, 40,
          (int)weather.getTemperature(), 20,
      a -> {
        tree.calculateSkyColor(((JSlider)a.getSource()).getValue());
      });

      controlFrame.addSeperator();
```

And here is our final, reworked Canvas class, which can also be found on the book's website.

Listing 4.10 Final Canvas class

```
1  import java.awt.Dimension;
2
3  import javax.swing.BoxLayout;
4  import javax.swing.JComponent;
5  import javax.swing.JFrame;
6  import javax.swing.JLayeredPane;
7  import javax.swing.JPanel;
8  import javax.swing.JSlider;
9  import javax.swing.JToggleButton;
10 import javax.swing.Timer;
11
12 class Canvas extends JPanel {
13   private JLayeredPane layeredPane;
14   private Timer timer;
15
16   public Canvas() {
17     setLayout(new BoxLayout(this, BoxLayout.PAGE_AXIS));
18     layeredPane = new JLayeredPane();
19         layeredPane.setPreferredSize(new Dimension(600, 600));
20
21         Tree tree = new Tree();
22         Rain rain = new Rain(600);
23
24     OpenWeather weather = new OpenWeather();
25     Timer weatherRefresh = new Timer(600000, a -> { //every 10 minutes
26       weather.requestWeather();
27       tree.calculateSkyColor(weather.getTemperature());
28       tree.calculateSwayBias(weather.getWindDirection());
29       rain.setWind(weather.getWindDirection());
30       rain.setVisible(weather.isRaining());
31     });
32
33     tree.calculateSkyColor(weather.getTemperature());
34     tree.calculateSwayBias(weather.getWindDirection());
35     rain.setWind(weather.getWindDirection());
36     rain.setVisible(weather.isRaining());
37
38     weatherRefresh.start();
39
40     tree.setBounds(0, 0, 600, 600);
41     rain.setOpaque(false);
42     rain.setBounds(0, 0, 600, 600);
43     layeredPane.add(rain);
44     layeredPane.add(tree);
45     add(layeredPane);
46
47     timer = new Timer(100, a -> layeredPane.repaint());
48     timer.start();
49
50     //Control Frame for override controls
51     ControlFrame controlFrame= new ControlFrame();
52     initControlComponents(controlFrame, tree, weather, rain);
53     controlFrame.setBounds(1250, 50, 300, 600);
54     controlFrame.setVisible(true);
55   }
56
57   public static void createAndShowGUI() {
58     JFrame frame= new JFrame();
59     frame.setDefaultCloseOperation(JFrame.EXIT_ON_CLOSE);
60     frame.setLocation(600, 50);
61     frame.setResizable(false);
```

```
62
63      JComponent newContentPane = new Canvas();
64          newContentPane.setOpaque(true);
65          frame.setContentPane(newContentPane);
66
67      frame.pack();
68      frame.setVisible(true);
69    }
70
71    public static void main(String[] args) {
72      javax.swing.SwingUtilities.invokeLater(new Runnable() {
73              public void run() {
74                  createAndShowGUI();
75              }
76          });
77    }
78
79
80    public static void initControlComponents(ControlFrame controlFrame, Tree
          tree, OpenWeather weather, Rain rain){
81      controlFrame.addHeading("City Search");
82      controlFrame.addSubmissionBox("Set City", weather, tree, rain);
83      controlFrame.addButtonControl("Print Weather To Console", a ->
            System.out.println(weather));
84
85      controlFrame.addSeperator();
86
87      controlFrame.addHeading("Weather Controls");
88      controlFrame.addSlideControl("Wind Direction", 0, 360,
            (int)weather.getWindDirection(), 45, a -> {
89        rain.setWind(((JSlider)a.getSource()).getValue());
90        tree.calculateSwayBias(((JSlider)a.getSource()).getValue());
91        tree.repaint();
92      });
93      controlFrame.addCheckBoxControl("Rain Toggle", weather.isRaining(), a
            -> {
94        rain.setVisible(((JToggleButton)a.getSource()).isSelected());
95        tree.repaint();
96      });
97
98      controlFrame.addSlideControl("Rain Amount", 0, 1000, 600, 200, a -> {
99        rain.setDrops(((JSlider)a.getSource()).getValue());
100       tree.repaint();
101     });
102
103     controlFrame.addSlideControl("Temperature (C)", 0, 40,
            (int)weather.getTemperature(), 20, a -> {
104       tree.calculateSkyColor(((JSlider)a.getSource()).getValue());
105     });
106
107     controlFrame.addSeperator();
108
109     controlFrame.addHeading("Tree Controls");
110     controlFrame.addSlideControl("Width", 0, 100, tree.width,  a -> {
111       tree.width = ((JSlider)a.getSource()).getValue();
112       tree.repaint();
113     });
114
115     controlFrame.addSlideControl("Length", 0, 200, tree.length, a -> {
116       tree.length = ((JSlider)a.getSource()).getValue();
117       tree.repaint();
118     });
119     controlFrame.addSlideControl("X", 0, 600, tree.x, a -> {
120       tree.x = ((JSlider)a.getSource()).getValue();
121       tree.repaint();
122     });
123     controlFrame.addSlideControl("Y", 0, 600, tree.y, a -> {
```

```
124         tree.y = ((JSlider)a.getSource()).getValue();
125         tree.repaint();
126       });
127       controlFrame.addSlideControl("Length Reduction", 0, 10,
              tree.lReduction, 1, a -> {
128         tree.lReduction = ((JSlider)a.getSource()).getValue();
129         tree.repaint();
130       });
131       controlFrame.addSlideControl("Width Reduction", 0, 10,
              tree.wReduction, 1, a -> {
132         tree.wReduction = ((JSlider)a.getSource()).getValue();
133         tree.repaint();
134       });
135       controlFrame.addSlideControl("Branching Number", 0, 10,
              tree.branchingNum, 1, a -> {
136         tree.branchingNum = ((JSlider)a.getSource()).getValue();
137         tree.repaint();
138       });
139       controlFrame.addSlideControl("Angle", 0, 180, tree.angle, 45, a -> {
140         tree.angle = ((JSlider)a.getSource()).getValue();
141         tree.repaint();
142       });
143
144       controlFrame.addButtonControl("Default Values", a ->
              tree.resetTreeValues());
145
146
147
148     }
149   }
```

But, before we can compile this we need to add our **Rain** class and make some changes to the **Tree** class also.

4.3.7 Adding rain to our display

Apart from rain being a common weather event phenomenon, and so should be included in our weather data display app, it can also serve as an indicator of wind direction. We can simply draw short length lines to represent our rain. But, like in the real world, rain does not fall at a regular rate. The drops of rain appear to fall in different sizes and at different rates. To generate this effect we will need to use a random number generator. This will randomly vary the length of our rain drop lines and their intervals across our display. Finally we need to adjust the angle and direction from which they appear to enter our display.

From the code you will notice we need to `import java.util.Random`. This is the standard Java method in the `utils` class for generating random numbers. Next we need to draw our rain drop lines onto a panel. This will be layered on top of our tree panel in the final `Canvas` window or frame. Our variables are declared (`dropsXY, dropSpeeds, random, MIN_SPEED`, and `wind`) before our main `Rain` method is initialised. Next, the number of drops is defined. The `setDrops()` method sets up random locations for the first raindrops and random speeds (within a range) for the number of drops asked for. The panel is set to the preferred size 600, 600 to match the tree panel.

`Wind` is initially set at 0 then updated from the canvas class which gets its information from the `openweather` class; the wind value is in degrees from 0 to 360. The value assigned to wind is used as the angle needed to rotate the rain before it is painted to the panel. The angle between 340-45 degrees is mirrored if the wind is between 135-225 degrees and clamps the direction at the extremes between them all – otherwise the rain could be horizontal. The `paintComponent()` method is called at the rate set by the timer. It is used to rotate the graphics by the wind amount – 300, 300 is at the centre of our 600, 600 panel and thus the centre pivot for our rotation. Colour is set to blue and then for each of the drops it is calculated whether they will move off the screen (> 700) after moving their set amount. If they do move off the screen (which they must), they are reset to a position above the panel in a random `x` position.

```
dropsXY[i] = random.nextInt(800) - 100;
```

This widens the `x` bounds of where drops can spawn. It helps ensure drops cover the screen when rotating so they can fall again. Lines are then drawn in their spots at a length determined by their speed.

Listing 4.11 Rain

```
1  import java.awt.Color;
2  import java.awt.Dimension;
3  import java.awt.Graphics;
4  import java.awt.Graphics2D;
5  import java.util.Random;
6
7  import javax.swing.JPanel;
8
9  public class Rain extends JPanel {
10
11   private int [] dropsXY;
12   private int [] dropSpeeds;
13   private Random random = new Random();
14   private final int MIN_SPEED = 20;
15   private double wind;
16
17   public Rain(int amount){
18     setDrops(amount);
19     setPreferredSize(new Dimension(600, 600));
20   }
21
22   public void setDrops(int amount) {
23     dropsXY = new int [amount*2];
24     dropSpeeds = new int [amount];
25     for (int i = 0; i < dropsXY.length; i++){
26       dropsXY[i] = random.nextInt(600);
27     }
28     for (int i = 0; i < amount; i++) {
29       dropSpeeds[i] = random.nextInt(20) + MIN_SPEED;
30     }
31   }
32
33   public void setWind(double wind){
34     if (wind > 225 & wind < 315)
```

```
35          this.wind = 315;
36        else if (wind > 45 & wind < 135)
37          this.wind = 45;
38        else if (wind >= 135 & wind <= 180)
39          this.wind = 180 - wind;
40        else if (wind > 180 & wind < 315)
41          this.wind = 540 - wind;
42        else
43          this.wind = wind;
44      }
45
46      @Override
47      protected void paintComponent(Graphics g) {
48        super.paintComponent(g);
49        ((Graphics2D)g).rotate(Math.toRadians(wind),300,300);
50        g.setColor(Color.BLUE);
51        for(int i = 0; i < dropsXY.length; i+=2){
52          if ((dropsXY[i+1] += dropSpeeds[i/2]) > 700){
53            dropsXY[i+1] = -100;
54            dropsXY[i] = random.nextInt(800) - 100;
55          }
56
57          g.drawLine(dropsXY[i], dropsXY[i+1], dropsXY[i], dropsXY[i+1] -
                (dropSpeeds[i/2])/2);
58        }
59      }
60
61    }
```

4.3.8 Driving the Tree class from the weather data

After building our basic Tree graphic, we added a control panel which included sliders so we could manipulate its elements – branches, position, size and so on. Now we can use those same parameter controls as inputs for data from our openweather data feed. To do this we need to declare some new variables that can be accessed from the OpenWeather class. This involves using the wind direction data to influence the sway of our tree – making it move in the direction that the wind pulls it. We can also use the minimum and maximum daily temperature ranges to define a colour gradient for our background.

In order to use the information from our OpenWeather class we need to declare our variables – sway, sway, swayUpBound, swayLoBound, swayHeadingUp (the full class code is included in the book's website). Next we specify our background colour method with some default values.

```
GradientPaint backgroundColor = new GradientPaint(0, 0, Color.BLACK, 0, 0,
     Color.WHITE);
```

Inside our paintComponent() method (which is a child of the super paint method) where we build the background we need to add also our tree graphic, that it will be swayed, and a pot for its base.

```
1  public void paintComponent(Graphics g){
2    super.paintComponent(g);
3    paintBackground((Graphics2D)g);
4    adjustSway();
5    drawPot(g);
6    drawTree((Graphics2D)g);
7    g.setColor(Color.BLACK);
```

Next we want to fill a rectangle with our background colours and combine them using a colour gradient between the top and bottom of the display panel.

```
1  private void paintBackground(Graphics2D g) {
2    g.setPaint(backgroundColor);
3    g.fillRect(0, 0, getWidth(), getHeight());
4  }
5
6  public void setBackgroundColor(GradientPaint backgroundColor) {
7    this.backgroundColor = backgroundColor;
8  }
```

We need a method to execute this function:

```
1  public void calculateSkyColor(double temperature) {
2    int maxTemp = 40;
3    int r = (int) (Math.sin(Math.toRadians(temperature * 9 + 90)) * 127.5 +
         127.5);
4    int b = 255 - (int)((temperature/maxTemp)*255);
5    int g = 255 - (int)((temperature/maxTemp)*255);
6
7
8    //Brighten Colour
9    double brightness = (Math.sin(Math.toRadians(temperature * 5 - 90)) *
         0.2 + 0.2);
10   r = (int) Math.round(Math.min(255, r + 255 * brightness));
11   g = (int) Math.round(Math.min(255, g + 255 * brightness));
12   b = (int) Math.round(Math.min(255, b + 255 * brightness));
13
14   Color c1 = new Color(r, g, b);
15   Color c2 = new Color(r, 255, b);
16   setBackgroundColor(new GradientPaint(300, 400, c1, 300, 420, c2));
17 }
```

The pot we are using as a base for our tree needs to be drawn and located also. It is constructed using a series of filed ovals and rectangles. Two colours are used, grey and dark grey. The ovals are layered such that they form the top of the pot and its bottom. Rectangles are used to bridge between the ovals filling in the remainder of the pot.

```
1  private void drawPot(Graphics g) {
2    g.setColor(Color.GRAY);
3    int potX = x-width*2;
4    int potY = y+(int)(length*1.9);
5    int potW = width*5;
6    int potH = (int)(width);
7    g.fillOval(potX, potY, potW, potH);
8    g.fillRect(potX, potY+(potH)/2, potW, potH);
```

```
9      g.fillOval(potX, (potY+(potH)/2) + (potH/2), potW, potH);
10     g.setColor(Color.DARK_GRAY);
11     g.fillOval(potX+(int)(potW * 0.1), potY+(int)(potH * 0.1), (int)(potW *
           0.8),
12     (int)(potH * 0.8));
13     g.setColor(Color.GRAY);
14     g.fillRect(potX+(int)(potW * 0.1), (potY+(potH)/2) + (potH/2) + (potH/2),
15     (int)(potW * 0.82), potH*2);
16     g.fillOval(potX+(int)(potW * 0.1), (potY+(potH)/2) + (potH/2) + (potH/2)
           +
17     ((potH*2)-(potH/2)), (int)(potW * 0.82), potH);
18   }
```

When our tree gets drawn we need to add the sway functionality. To do this we need to add the following statements. These are used to affect the width, length and angle of the trunk and `trunkBottom` of the tree:

```
1   void drawTree(Graphics2D g2d){
2     ...
3     g2d.rotate(Math.toRadians(sway + swayBias/1.2), x+width/2, y+length*2);
4     g2d.rotate(Math.toRadians(sway + swayBias), x+width/2, y);
5     double angle = this.angle - Math.abs(swayBias/2);
6     ...
7   }
```

Next we need methods for utilising the wind information from the `OpenWeather` class to influence our tree components. We have two methods for this – `calculateSwayBias` and `adjustSway`. The first calculates how much the wind data will influence the tree elements and in which direction from side to side. The second checks whether the direction is forward or aft.

```
1   void calculateSwayBias(double wind){
2     if (wind >= 45 & wind < 135)
3     wind = -45;
4     else if (wind >= 135 & wind <= 225)
5     wind = wind - 180;
6     else if (wind > 225 & wind <= 315)
7     wind = 45;
8     else if (wind > 315 & wind <= 360)
9     wind = 360 - wind;
10    else wind = -wind;
11    swayBias = wind/4;
12  }
13
14  private void adjustSway(){
15    if(swayHeadingUp){
16      sway+=0.1;
17      if(sway >= swayUpBound){
18        swayHeadingUp = false;
19      }
20    }else{
21      sway-=0.1;
22      if(sway <= swayLoBound){
23        swayHeadingUp = true;
24    }
25  }
```

And finally we need to add our sway variables to our reset values so we can get back to the default display after using the sliders to simulate effects.

```
1  void resetTreeValues(){
2    ...
3    sway = 0;
4    swayBias = 0;
5    swayUpBound = 0.5f;
6    swayLoBound = -0.5f;
7    ...
8  }
```

The final, reworked `Tree` class can be found on the book's website.

4.3.9 Adding features to the control panel

Because all we are doing is accessing some of the control features in our `ControlFrame` class, substituting their values with those from our `OpenWeather` class data, it remains largely unchanged. However, we need to add those features unique to the `OpenWeather` class – the ability to specify a city, simulate rain and a button to print weather data to the console. Each of these controls' parameters needs to be specified in the control panel.

```
1  public void addCheckBoxControl(String label, boolean value,ActionListener
       al){
2    JCheckBox checkbox = new JCheckBox(label, value);
3    checkbox.addActionListener(al);
4    checkbox.setBackground(new Color(225, 225, 250));
5    checkbox.setAlignmentX(LEFT_ALIGNMENT);
6    container.add(checkbox);
7    container.add(Box.createVerticalStrut(15));
8    pack();
9  }
10
11 public void addSubmissionBox(String label, Tree, OpenWeather weather, Tree
       tree, Rain rain){
12   JTextField textField = new JTextField();
13   JButton button = new JButton(label);
14   button.addActionListener(a -> {
15     weather.requestWeather(textField.getText());
16     tree.calculateSkyColor(weather.getTemperature());
17     tree.calculateSwayBias(weather.getWindDirection());
18     rain.setWind(weather.getWindDirection());
19     rain.setVisible(weather.isRaining());
20   });
21   JPanel boxPanel = new JPanel();
22   boxPanel.setLayout(new BoxLayout(boxPanel, BoxLayout.X_AXIS));
23   boxPanel.add(textField);
24   boxPanel.add(button);
25   boxPanel.setAlignmentX(LEFT_ALIGNMENT);
26   container.add(boxPanel);
27   container.add(Box.createVerticalStrut(15));
28   pack();
29 }
```

And in the book's website, you can find the final, reworked `ControlFrame` class.

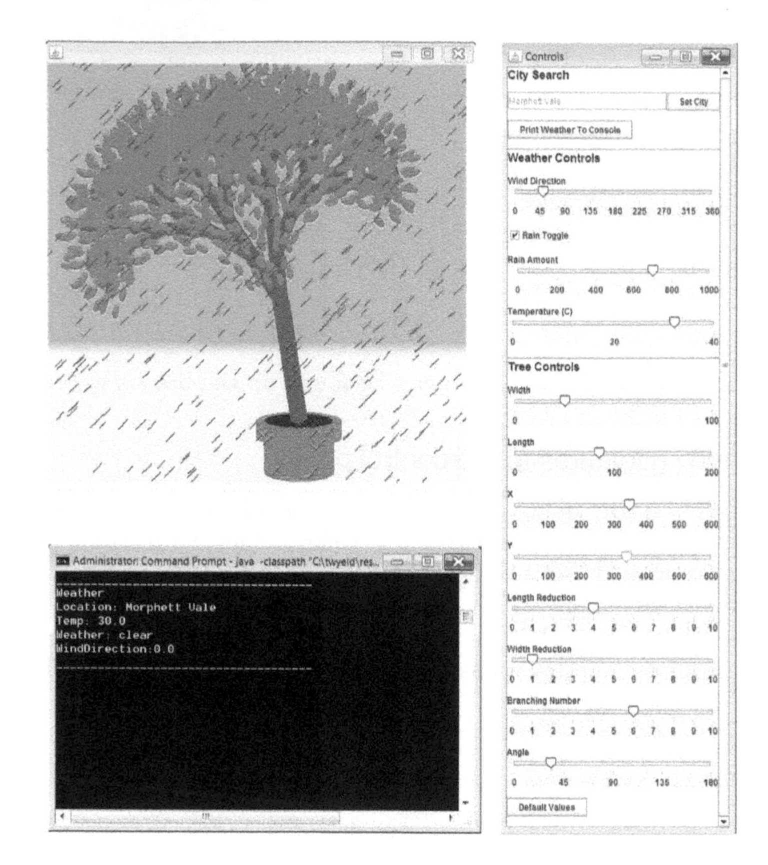

FIGURE 4.8 Weather Tree app showing data feed for Morphett Vale, SA, Australia. Note no rain and wind direction 0.0, but sliders used to simulate rain at 45^0 North. © Theodor Wyeld. Used with permission.

When all of these extra code snippets are added to the various class files, compiled and run you should see Figure 4.8; it may take a while to load, as it is trying to connect to the weather station[3].

4.4 ADDING BLING: USING LEDS WITH ARDUINO

By Erik Brunvand
The School of Computing, University of Utah

[3]You may need to adjust the security settings on your PC if you have trouble connecting – it needs to establish a connection on a specific port and URL address.

4.4.1 Introduction

Having learnt the basics of programming and some creative thinking, you may want to explore working on an interactive project with hardware. Arduino is an open source prototyping platform used for interactive projects based on hardware and software, with boards that read input from a sensor, a finger, a button, or a Twitter message and turns it into an output that activates a motor, turns on an LED, or publishes something online.

This section shows how you can add physical lights to your work using light emitting diodes (LEDs). Of course, because these are physical lights, this makes sense mostly for work that you have printed out on paper or other physical materials, not for on-screen images. LEDs are semiconductor components that light up when you pass a small amount of electrical current through them. They are easy to control if you have a suitable "physical computing" platform. Physical computing means that your program (the "computing") controls something in the physical world such as turning an LED on and off. This section will introduce the Arduino physical computing platform, the Arduino programming language (a variant of the C language), some background on the electronics required, and some examples of simple programs that control how and when LEDs light up.

4.4.2 Arduino

Controlling physical things such as LEDs requires a computing platform that is designed to connect your program to physical electrical components. Arduino is one such platform that is designed specifically to do this. It is a small, inexpensive, and widely used computing platform for exactly this purpose. This lesson is specifically about controlling LEDs, but the Arduino can also control motors and servos for motion, and receive input from a wide variety of sensors such as switches, light sensors, temperature sensors, and many many more.

4.4.3 Downloading and installation

Information about all things Arduino can be found on the main Arduino website: `https://www.arduino.cc`. To follow along with this lesson, you will need the following resources.

You will need to acquire an Arduino board. There are many different shapes and sizes of Arduino, and they all do pretty much the same thing. The generic Arduino is a very simple credit card sized computer that costs around $20 or £15. In fact, Arduino is "open souce" both in the traditional software sense, but also in the hardware sense meaning that anyone can build and sell something based on the Arduino circuit. So, perfectly reasonable versions of Arduino can be found on foreign websites for as little as $5.

There are many different sizes and shapes of Arduino. To start with, you might want to use the very generic Arduino UNO seen in Figure 4.9. These are available through a wide variety of web resources such as “maker” sites like Sparkfun (`sparkfun.com`) and Adafruit (`adafruit.com`), or directly from the Arduino website at `store.arduino.cc`.

Once you have an Arduino board, you will need the Arduino software environment on your desktop or laptop computer in order to write programs and upload and run them on your Arduino board. This software environment is free and open source, and runs on Mac, PC, and Linux. To download, go to the Arduino software page (`https://www.arduino.cc/en/Main/Software`) and download the integrated development environment (IDE) for your computer type (Mac, PC, or Linux). This is the environment that you will use to write programs for Arduino, compile those programs into machine instructions that the Arduino board understands, and upload those programs onto your Arduino board.

To make things light up you will need some LEDs, a few resistors, some wire, and a solderless breadboard to start with. You can acquire these from any electronics shop, but perhaps an easier approach to start with is to buy an Arduino starter kit that includes an Arduino board, and simple electronics parts. These starter kits can be obtained at any of the sites mentioned in the previous bullet. A search for “Arduino starter kit” returns a number of interesting kits. You can also find a starter kit on the main Arduino website at `https://store.arduino.cc/usa/arduino-starter-kit`.

4.4.3.1 Electronics basics

The part of electronics we are interested in for the purpose of lighting up LEDs is electrical current moving through a conductor. Current is measured in Amperes (usually shortened to Amps), and moves through a conductor (like a wire) under the influence of a driving force measured in Volts. That is, voltage is the electromotive force that causes current to flow. More voltage means more current flowing. The quality of the conductor also influences how much current flows with a given voltage. Some materials are great conductors (like copper), and some are only so-so (like graphite). The measure of the quality of the conductor is called resistance and is measured in Ohms (using the symbol Ω).

There is a simple formula that relates all of these quantities called Ohm’s Law. Ohm’s Law tells us that there is a direct relationship between current (Amps), electromotive force (Volts) and resistance (Ohms):

$$V = IR \tag{4.1}$$

FIGURE 4.9 Arduino UNO microcontroller. This small board (2.7 in × 2.1 in) has a microcontroller chip, a USB interface for uploading programs, and a wide variety of controllable pins for physical computing applications. Digital pins are at the top of this figure numbered 0 through 13. They can be used to connect and control LEDs from a program running on the Arduino. © Erik Brunvand. Used with permission.

In this formula, Voltage is V, current is (somewhat confusingly) I, and resistance is R. This formula says that voltage equals the product of current and resistance. This is a remarkable relationship and one that makes intuitive sense as well. One way to wrap your head around this is to think of electrical current as water, and the conductor as a pipe. The resistance of the conductor is like the diameter of the pipe – smaller diameter is higher resistance, larger diameter is less resistance. Voltage is like the pressure forcing the water through the pipe.

Using this analogy, if you have a constant water pressure (voltage), and make the pipe smaller (more resistance), then less water (current) will flow because of the higher resistance. Another intuition is that if you have a fixed sized pipe (fixed resistance) and you want more water to flow (more current), you will need to add more pressure (voltage) to make it all work out. Using Ohm's Law, if you know any two of the quantities (voltage, current, or resistance) you can solve for the third. Very handy indeed!

For our purposes we will want to understand how much current we are putting through an LED: Too little and the LED will not light up, too much and the LED will burn out. So, we will need to carefully control the voltage and resistance to result in just the right amount of current. In practice, for our purposes, the voltage will usually be fixed at 5v, and we will want a specific amount of current for the LED (0.018 amps, for example), so we will be solving for the correct amount of resistance to make that happen (i.e. $R = V/I$). More about this in the section about wiring up the LEDs.

4.4.4 Arduino programming

The Arduino is programmed in C/C++ through a nice integrated development environment (IDE) available at `www.arduino.cc/en/Main/Software`. An IDE is basically a wrapper around the programs required for developing code for the Arduino. It includes an editor, lots of great library functions, a compiler, and an uploading system to load your program onto the Arduino board. It also comes with lots of great example code to get you started. In "Arduino-speak" they call programs "sketches" to make them sound less scary, but I prefer just to call them programs or code.

The absolutely simplest Arduino program is "Blink." It comes with the IDE in the Examples-Basics menu tab. Figure 4.10 is a screen shot of the Arduino IDE with the Blink program loaded. This is a great program to start with for two reasons – it demonstrates a lot of issues related to coding in C/C++, and it flashes an LED and that is what we actually want to do. In this case there is no external wiring required because every Arduino UNO comes with an LED on the board that is already connected to pin 13 on the board. That way you can test your board by loading the Blink program (sketch) and make sure things are working.

Looking at this code there are a variety of things to notice:

- There are two forms of comments in the C/C++ language. You can start a comment with // and the comment goes to the end of the current line, or you can surround code in /* ... */ and everything between the /* and the */ will be considered a comment (it does not get computed). You can see both types of comments in Figure 4.10. The IDE editor recognises comments and turns comment text grey.
- All Arduino programs have two required functions. One is called `setup()` and the other `loop()`. Every program you write for Arduino must have these two functions. They both return nothing when they are called – that is what the `void` qualifier means. They also take no arguments so the function argument list is empty. You can see in Figure 4.10 that the IDE editor colour codes things when it can determine their function: Keywords are in blue, required functions in brown, comments in grey, and library functions in orange.
- The two required functions are as follows. Note that the code that makes up the function body is between the { ... } characters.

```
1  void setup() {
2    // code in the setup() function is executed exactly once when the
3    // program starts running. You can use it to initialize things
4  }
```

FIGURE 4.10 A screenshot of the Arduino IDE with the Blink program loaded. © Erik Brunvand. Used with permission.

```
1  void loop() {
2    // After setup is finished, the loop function is called. When it's
3    // finished it's called again, forever. It's where you put your
4    // action code, like the code to make the LED blink
5  }
```

- Library functions are provided by the Arduino IDE to help make writing physical computing code easier. Library functions are coloured orange in the IDE. Pre-defined keywords are in blue. Note that every line of code in the body of the functions must be ended with a semicolon. The library functions in Figure 4.10 are:

  – `pinMode(pinNumber, Direction);`

    This function sets the direction (`INPUT` or `OUTPUT`) of one of the 14 digital pins on the Arduino board. You can use any of these pins (numbered 0 through 13) to drive your own LEDs, but if you do, you must first set them as `OUTPUT` pins meaning that the Arduino will drive the voltages on those pins. The other option is `INPUT` where

FIGURE 4.11 Control icons. © Erik Brunvand. Used with permission.

you sense an external voltage on that pin, but we will not use that mode here. In Figure 4.10 the `LED_BUILTIN` keyword evaluates to pin 13. You could have written this line with a 13 instead.

- `digitalWrite(pinNumber, Voltage);`

  This sets the voltage on the pin to be the pre-defined values of `HIGH` (+5 volts) or `LOW` (0v). This is the primary mechanism for using Arduino to control physical things. You can use it to turn LEDs on and off, turn motors on and off, etc.

- `delay(milliseconds);`

  This stops the program for the given number of milliseconds. A millisecond is a thousandth of a second, so to make the program delay for 1 sec, you give it an argument of 1,000 milliseconds (1,000 ms).

- You can get much more information on all the library functions that the Arduino IDE offers on the Arduino website at `arduino.cc/reference/en`.

• The control icons at the top of the editor window (see Figure 4.11) let you evaluate your code. They are (from left to right):

  - Verify: Compile your code and see if there are any errors – errors will show up in red in the bottom of the window
  - Upload: Compile your code and upload it to the Arduino board connected through the USB cable
  - New: Open new blank code template that includes the setup() and loop() functions
  - Open: Open an existing program
  - Save: Save your code

Knowing these things should now make it easier to understand what this program does:

1. First `setup()` runs and defines the `LED_BUILTIN pin` (pin 13) to be an `OUTPUT` pin
2. Then `loop()` runs

   (a) First it sets the voltage on pin 13 to `HIGH`, turning on the built-in LED
   (b) Then the program waits for 1 sec (1,000 milliseconds)
   (c) Then it sets the voltage on pin 13 to `LOW`, turning off the LED
   (d) Then it waits for another second

3. Then `loop()` runs again. And again. And again. Until the Arduino is disconnected.

The effect is that the built-in LED will flash on for 1 second, then off for 1 second, and repeat until you disconnect the Arduino. A very easy change is to modify the delay timing and see what effect that has on the blinking rate.

4.4.4.1 *Uploading this code to your Arduino board*

When you connect your Arduino UNO board to your machine, it should inform your operating system that it is connected. Once it is connected you should go to the Tools menu in the Arduino IDE and make sure that you select the correct board ("Arduino/Genuino Uno" if you are using a board like that in Figure 4.9), and the correct port (the port that has Arduino in the port name). Now the IDE knows what code to upload and to where. When you press the upload button (the right-arrow), your code will compile and upload to the board. If you watch closely, you will see a pair of LEDs on the Arduino board flicker as the transfer is made.

4.4.5 Wiring LEDs to Arduino

A light emitting diode (LED) is a semiconductor component that lights up when current is passed through it. It is a version of a more generic component called a diode that serves as a one-way valve for current. That is, current can pass through a diode from input to output (anode to cathode), but not backwards. It is like an anti-backflow valve in a plumbing system if you want to continue the water/electricity analogy. A light emitting diode is a diode that lights up when there is current flowing from input (anode) to output (cathode). To make current flow you make the voltage at the anode higher than the voltage at the cathode: Current flows from a higher voltage to a lower voltage. The amount of current can be computed using Ohm's Law for a given voltage and resistance.

LEDs are now used in energy efficient lighting because they take very little current to make them light up. They are much more energy efficient per unit of light emitted than an old-fashioned incandescent bulb. But, they are sensitive to the amount of current passing through them. A generic LED like those in Figure 4.12 will light up when between 10mA and 20mA flow through the device. A mA is a "milliamp" or thousandth of an amp. 10mA is the same as 0.010 amps.

FIGURE 4.12 Variety of different LEDs. All LEDs have two connections: An input called the anode and an output called the cathode. When current flows from anode to cathode, the LED lights up. © Erik Brunvand. Used with permission.

To make sure that the current is limited to that amount, we can use Ohm's Law and the given voltage to compute the current, and add a current-limiting resistor if needed. A resistor is an electrical component that adds resistance to a circuit. When you add resistance to a circuit with a constant voltage, Ohm's Law tells us that less current will flow. Figure 4.13 shows some resistors. The ones you will likely be using are the striped cylinders at the top of the figure. The stripes are used as a code to tell you the value of the resistance. It is a tricky code to figure out, so I like to use an on-line resistor calculator such as `www.hobby-hour.com/electronics/resistorcalculator.php`. There are lots of them out there!

4.4.5.1 Practicalities of wiring and powering circuits

Circuits can be wired up using any conductive material, but insulated wires are the easiest and most common materials used. A schematic is a diagram that tells you about the logical organisation of the electrical components. It uses standard symbols that represent electrical components and lines that represent electrical nodes. It is a recipe for how you should assemble the physical circuit.

The physical realisation of that circuit could look very different, but as long as you connect the components into the same set of components as in

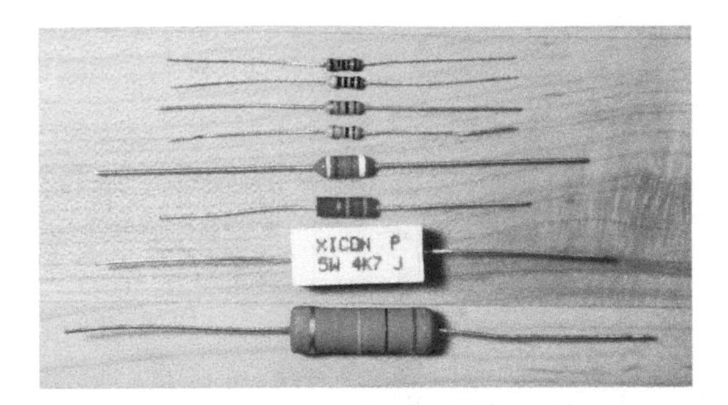

FIGURE 4.13 Some example resistors. Generic small resistors are shown at the top and are likely what you will mostly see. The larger resistors are for higher-power applications. © Erik Brunvand. Used with permission.

the schematic, it will have the same function as the schematic. For example, Figure 4.14 is an extremely simple schematic that describes two components connected in series: An LED (the triangle-ish component) and a resistor (the zig-zag). The schematic says that you should connect one terminal of the resistor to Pin 10 of the "Arduino Controller", and the other end of the resistor to one end of the LED (the anode). Finally, the other end of the LED (the cathode) should be connected to GND (ground, or 0 volts).

Physically, LEDs come in a wide variety of shapes, sizes, and colours. One important consideration is how to figure out which lead is the anode and which is the cathode. In a standard through-hole LED (i.e. one that is designed to be inserted through the holes in a circuit board), one lead will be longer than the other. The longer lead is the anode, and the shorter is the cathode. The good news is that unless you apply a very large voltage to the leads of the LED you cannot really hurt it by putting it in a circuit backwards. That is what diodes are designed for after all, to block current in the reverse direction. So, if you do not remember which lead is which, or if you have cut the leads of the LED to the same length, you can just try it both ways and see which way lights up.

Resistors are not directional. There is no input or output side of a resistor and it can be inserted in either orientation without changing the function.

To make this circuit you need to find physical examples of all the main components (the "Arduino Controller", the resistor, and the LED) and use wires to connect them together. This schematic doesn't specify the resistance value of the resistor, or the colour or electrical specifications of the LED. In a real circuit that information would be annotated on the schematic, or in a supplemental document that references the component names R1 and D1. You

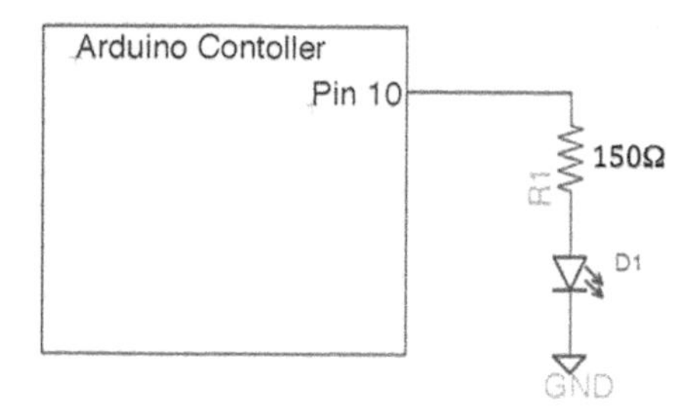

FIGURE 4.14 Simple schematic diagram showing an Arduino connected through Pin 10 to a resistor in series with an LED to ground. © Erik Brunvand. Used with permission.

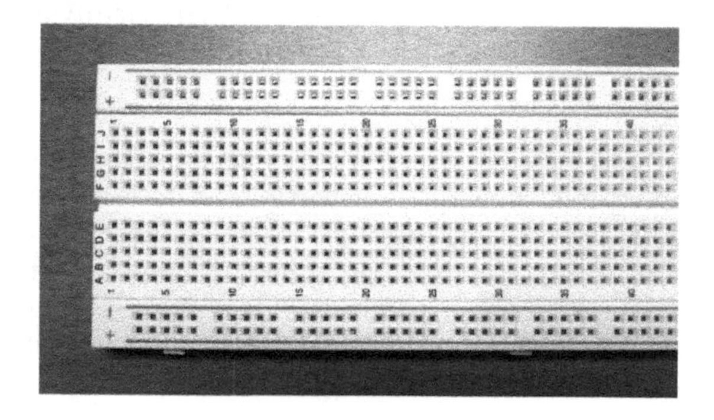

FIGURE 4.15 Solderless breadboard. To make connections, 22AWG solid core wire can be inserted into the holes. Component leads from LEDs and resistors can also be directly inserted into the holes to make connections. © Erik Brunvand. Used with permission.

could solder wires and components together to make a nice solid permanent bond, and in fact you probably want to do this for your finished product. But, for prototyping and testing, it is nice to have a less permanent, and less complex solution.

One way of doing this is with a "solderless breadboard" or "prototyping board" (see Figure 4.15). This is a board that has lots of little holes in it into which you can poke a wire. Inside the board are connections such that all the wires plugged into one row will be electrically connected. Using these breadboards, you can prototype a circuit by plugging and unplugging wires and components into the holes, and you can quickly change things to try out new ideas, or fix bugs.

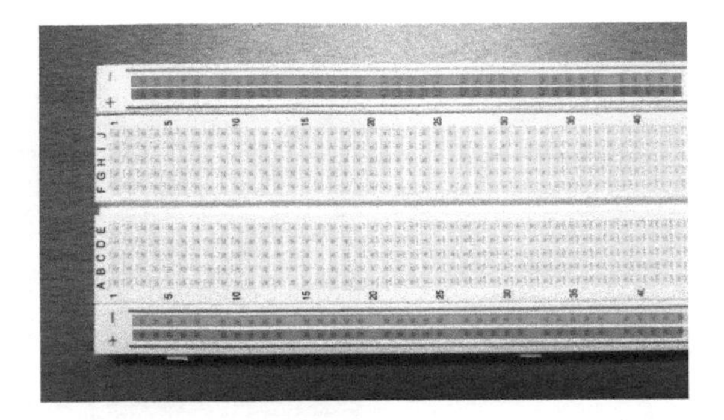

FIGURE 4.16 Solderless breadboard annotated to show what sets of pins are connected together internally to the breadboard. The holes in the rows at the top and bottom are connected into "buses." Every hole under the red and blue lines is connected internally. The sets of five holes in yellow are also connected. © Erik Brunvand. Used with permission.

In this breadboard certain rows and columns are connected inside the breadboard so that wires can connect by being plugged into the same row or column as another wire. Figure 4.16 shows how things are connected. The set of holes on either side of the board that are marked with red and blue lines are buses. Any wire plugged into the holes along the coloured line will be connected to any other wire plugged into that column. The entire red or blue column is one electrical node. These columns are typically (but not always) used for power and ground connections, so they are marked with + and – in case you want to use them for that purpose.

In this board, the five holes in a row marked with ABCDE are also connected as a node. That is, any wire plugged into row 5, column A will be connected to any other wire in columns B, C, D, or E of that row. Likewise, the F, G, H, I , and J holes are also connected in a row. There is no connection across the "valley" in the middle of the board. This is designed so that integrated circuit packages (chips) can be placed in the board with "legs" on either side of the valley. All sorts of other components can also be placed in these breadboards.

As a very simple example, here is the same circuit we saw before in Figure 4.14 with the resistor and LED. In the physical version of this circuit there is a purple wire connecting pin 10 on the Arduino to row 8 column A on the breadboard. Also in row 8 is one leg of the resistor in column E. The other side of the resistor is in row 5, column E. Also in row 5 (column A) is the anode of the LED. The cathode is connected to the blue ground column. That blue column is connected through the green wire back to the ground

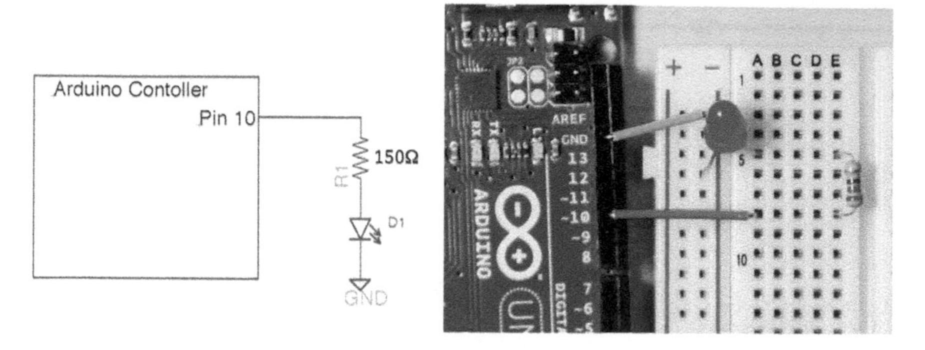

FIGURE 4.17 A schematic on the left, and the physical realisation of that schematic on the right using a solderless breadboard. © Erik Brunvand. Used with permission.

(GND) connection on the Arduino. Although it does not look identical, this is a correct physical realisation of the schematic using a solderless breadboard.

With the circuit wired up as in Figure 4.17, you could change the blink program to use pin 10 instead of 13 and it would flash the external LED. One way to do this would use a newly defined global variable named myLEDPin to hold the value of the pin number (10). It would look like the following:

```
int myLEDPin = 10; // define a global integer (int) variable to hold the pin number
void setup() {
  pinMode(myLEDPin, OUTPUT); // initialize digital pin 10 as an output.
}
void loop() {
  digitalWrite(myLEDPin, HIGH); // turn the LED on (HIGH is the voltage level)
  delay(1000); // wait for a second
  digitalWrite(myLEDPin, LOW); // turn the LED off by making the voltage LOW
  delay(1000); // wait for a second
} // loop() will repeat forever
```

What is going on at the circuit level is shown in Figure 4.18. When the Arduino pulls pin 10 to `HIGH` then there is a higher voltage (+5 volts) on the input of the LED than on the output (0 volts or ground). So, current will flow from the Arduino, through the resistor, through the LED, to ground. When the Arduino pulls the pin to `LOW`, then the input and output of the LED are at the same voltage, and because there is no voltage difference, no current will flow. By the way, because the components are connected in series, all the current must go through all the components. So, it does not matter if you wire the resistor in front of the LED, or the LED in front of the resistor. If they are in series, the order does not matter.

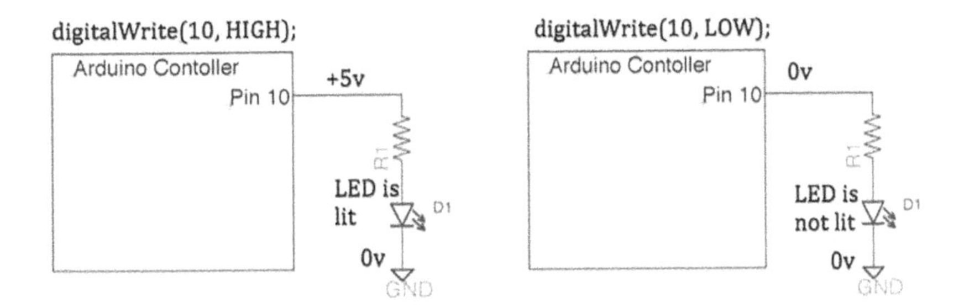

FIGURE 4.18 The electrical behaviour of the pin on the Arduino enables current to flow or not flow through the LED. © Erik Brunvand. Used with permission.

4.4.6 Current limiting resistor calculation

An important electrical consideration for any diode, LEDs included, is the "forward voltage." A diode does indeed conduct from anode to cathode, but only after the voltage difference from anode to cathode is raised above a certain voltage, known as the forward voltage, or Vf. For many regular non-lighting diodes, the forward voltage is 0.7v. For an LED, the forward voltage is higher: Often between 1.5v and 3.5v. Red LEDs typically have the lowest forward voltage (1.7v is a fairly typical value), and in general the forward voltage goes up as the frequency of the emitted light goes up. So, Vf goes up as LEDs move from up the spectrum (red, orange, yellow, green, blue) with blue LEDs having a Vf of 3.0 to 3.5v. Of course, the best plan is to get the specs for the LEDs you plan to use.

One ramification of the forward voltage is that the LED does not conduct until the voltage difference between anode and cathode gets above that value. Another is that because of that behaviour, the diode "uses up" that much voltage once it does conduct. That is, if you put 5v between anode and cathode of an LED with a 2.0v Vf, the voltage at the cathode will be 3v with respect to ground. Said in another way, the LED has a 2.0v voltage drop across the diode.

Another important consideration with respect to diodes is how much current the diode can support. Regular diodes come in a huge variety of current capacities from milliamps to hundreds or thousands of amps. Regular small LEDs, on the other hand, are almost always designed to produce maximum light output at around 20mA, and look good when the current is between 10-20mA. You can get super-bright high-powered LEDs that are designed for 800mA or 1000mA (i.e. up to 1A), but the generic inexpensive 3mm or 5mm LEDs that you are likely to encounter (e.g. Figure 4.12) will almost certainly be designed for a maximum current of around 20mA.

One question that is critical for using LEDs is how to make sure that the current you are using to drive the LED is limited to the right value. This is a perfect application for Ohm's Law. If we know the total voltage we are putting through the LED, the forward voltage Vf, and the desired current, we can use Ohm's law to compute the correct resistor value for that circuit.

Example: If the Arduino digital output pin drives to 5v, and the LED has a Vf of 2.0v, and a current limit of 20mA, then we can compute the resistor required to make this all work. Remember that the forward voltage Vf is subtracted from the total voltage ("used up" by the LED), and 20mA is 0.020A. Ohm's law uses whole-unit values for Ohms, Volts, and Amps.

$$V = IR, \text{ so } R = V/I, \qquad \text{Re-write the equation to solve for R} \tag{4.2}$$
$$R = Vtotal/I, \qquad \text{In our case, voltage is what's left after LED uses some up} \tag{4.3}$$
$$R = (Vsource - Vf)/I, \qquad \text{The total is the main voltage minus LED's forward voltage} \tag{4.4}$$
$$R = (5v - 2.0v)/0.020A, \qquad \text{Plug in the actual numbers for this example} \tag{4.5}$$
$$R = 150\Omega, \qquad \text{And the result is 150 ohms} \tag{4.6}$$

Therefore, in this example, you could make the resistor R1 in Figure 4.14, 8, and 9 a 150Ω resistor and that would result in 20mA of current. If you would rather be a little nicer to the LED and not push the maximum amount of current you could re-solve for, say, 15mA and use a 200Ω resistor. In practice, to be safe, and because sometimes the Vf is lower, a standard safe resistor to use when you do not know all the details is 330Ω or 470Ω. If your computation results in an awkward resistance value that you do not have on hand, size up to the next larger resistor that you have. It is always safer to use more resistance resulting in less current than the other way around.

Whether you use a safe general value, or compute the exact value, you should *always* use a current limiting resistor whenever you are connecting an LED in a circuit.

4.4.7 Controlling multiple LEDs

Of course, controlling one flashing LED is fine, but more is always better! So, how can you control multiple LEDs from your Arduino? There are many advanced ways of doing this that involve external integrated circuits designed for driving multiple LEDs. But, a simple way of controlling up to 14 LEDs is to connect one LED to each of the 14 digital pins on the Arduino. Thus, by setting the HIGH and LOW values of each of those pins, you can turn on and off each of the LEDs from your code.

When connected this way, each Arduino pin would have a copy of the circuit from Figure 4.14: One current limiting resistor and one LED on each pin. Figure 4.19 shows a schematic of a circuit that involves eight LEDs, each connected to a different digital pin on the Arduino.

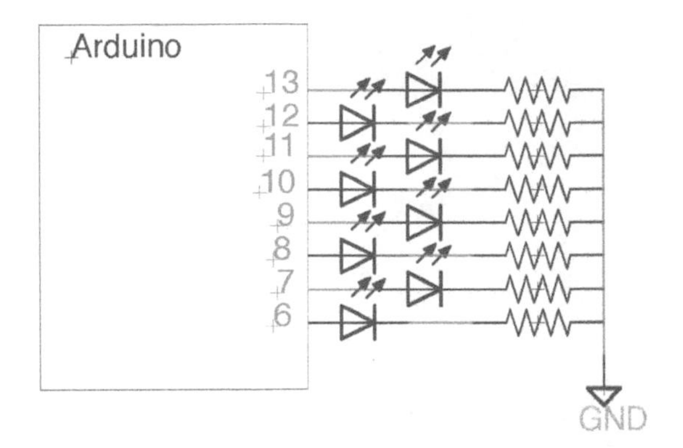

FIGURE 4.19 An example of eight LEDs connected to pins 6 through 13 of an Arduino. Each one can be individually controlled through the code by setting that pin to **HIGH** or **LOW**. Note that every LED needs its own current limiting resistor – you can't share that resistor among multiple LEDs. © Erik Brunvand. Used with permission.

There are many ways you could write an Arduino program to light up these LEDs. The easiest would be just to write individual **digitalWrite()** commands for each LED. That program would look something like the following. This is a brute-force approach and would take a lot of tedious repetitive coding to design a complex light pattern.

Listing 4.12 Eight LEDs: Light patterns

```
1  /*
2  Eight-LED example #1. This example takes a brute-force approach to making
       patterns appear on the 8 LEDs.
3  */
4  int delayMS = 100; // Variable to hold the time (in milliseconds) to delay
       between parts of the LED pattern.
5  void setup() { // Runs once to set up the pins as outputs
6    pinMode(6, OUTPUT);   // Define all 8 pins as outputs...
7    pinMode(7, OUTPUT);
8    pinMode(8, OUTPUT);
9    pinMode(9, OUTPUT);
10   pinMode(10, OUTPUT);
11   pinMode(11, OUTPUT);
12   pinMode(12, OUTPUT);
13   pinMode(13, OUTPUT);
14   digitalWrite(6, HIGH); // Start with the LED on pin 6 HIGH (ON)
15 }
16 void loop() {
17   // This set of digitalWrite() commands makes (half of) the "Cylon eye"
         behavior.
18   // Start by going from pin6 LED to pin13 LED.
19   // Note that the first time through loop(), pin6 is already HIGH
20   delay(delayMS);          // Wait for delayMS milliseconds
21   digitalWrite(6, LOW);    // set pin 2 LED OFF
22   digitalWrite(7, HIGH);   // set pin 3 LED ON
23   delay(delayMS);          // Wait for delayMS milliseconds
```

FIGURE 4.20 A wired circuit for the schematic in Figure 4.19. Note that each LED has its own current limiting resistor. Also, not seen in this photo, the blue column on the left is connected back to the ground (GND) pin on the Arduino. © Erik Brunvand. Used with permission.

```
24    digitalWrite(7, LOW);    // set pin 3 LED OFF
25    digitalWrite(8, HIGH);   // set pin 4 LED ON
26    delay(delayMS);          // Wait for delayMS milliseconds
27    digitalWrite(8, LOW);    // etc...
28    digitalWrite(9, HIGH);
29    delay(delayMS);
30    digitalWrite(9, LOW);
31    digitalWrite(10, HIGH);
32    delay(delayMS);
33    digitalWrite(10, LOW);
34    digitalWrite(11, HIGH);
35    delay(delayMS);
36    digitalWrite(11, LOW);
37    digitalWrite(12, HIGH);
38    delay(delayMS);
39    digitalWrite(12, LOW);
40    digitalWrite(13, HIGH);
41    delay(delayMS);
42    digitalWrite(13, LOW);
43    // For the full "Cylon eye" you would need to light up the LEDs
44    // in reverse order here...
45  }
```

A slightly fancier version of this program makes use of a separate function to set the state of all eight LEDs based on the value expressed as an 8-bit byte. This version makes use of several features of the C/C++ language used to program the Arduino that we have not seen yet:

- An array is used to hold the values of the pins to which the LEDs are connected. This allows the use of a loop to walk through the array to, for example, define all those pins as `OUTPUT`.

- It defines a new function called `setLEDs(byte,delay)` to light up a subset of the LEDs based on the bits in the argument byte.
- It uses the `B01010101` syntax to define an 8-bit byte of data.
- It uses another library function called `bitRead(byte, bit)` to extract a single bit from an 8-bit byte.

In this example, to fit on the page, the `setup()` function is in a separate figure from the `loop()`, and `setLEDs(byte,int)` functions. To run this you should put both code snippets into the same program in the Arduino IDE.

Listing 4.13 Cylon eye LED: setup()

```
1  /*
2  This example uses a separate function to set the LED outputs. This
       function can be called by the user each time the LED should be set.
       This example uses a relatively advanced low-level technique. It codes
       the LED values in a single byte with each bit of the byte being the 1
       or 0 that determines the LED ON/OFF state. This requires the
       setLEDs() function to pick off each bit of the byte in turn. */
3  /* Define the array to hold the LED pin numbers. It's defined here outside
       all the functions so that it's a "global" variable and can be seen by
       all functions. */
4  int delayMS = 100;  // Variable to hold the time (in milliseconds)
5  // to delay between parts of the LED pattern.
6  int ledPins[] = {6,7,8,9,10,11,12,13}; // An array to hold the Arduino
7  // pin numbers that each LED is connected to.
8
9  void setup() {
10   // Define all 8 pins as outputs...
11   // The pins are referenced through the ledPins array
12   for (int i = 0; i<8; i++){                    // loop eight times
13     pinMode(ledPins[i], OUTPUT);  // set each LED pin to OUTPUT
14   }
15 } // end of setup()
```

The Cylon eye is from Battlestar Galactica. The robots, called Cylons, have a light for an eye that moves back and forth in a scanning motion. It's the same scanning motion as used for "KITT" the AI-enhanced car in the 80s TV show, Knght Rider.

Listing 4.14 Cylon eye LED: loop() and setLEDs()

```
1  void loop() {
2    // This set of setLED() commands makes the "Cylon eye" behavior. This
         version takes a single byte as an argument to tell setLEDs what the
         ON/OFF state is in each step of the pattern. The Arduino syntax for
         this is B10101010 for a single byte value.
3    setLEDs(B00000001, delayMS);
4    setLEDs(B00000010, delayMS);
5    setLEDs(B00000100, delayMS);
6    setLEDs(B00001000, delayMS);
7    setLEDs(B00010000, delayMS);
8    setLEDs(B00100000, delayMS);
9    setLEDs(B01000000, delayMS);
10   setLEDs(B10000000, delayMS);
11   setLEDs(B01000000, delayMS);
12   setLEDs(B00100000, delayMS);
```

```
13      setLEDs(B00010000, delayMS);
14      setLEDs(B00001000, delayMS);
15      setLEDs(B00000100, delayMS);
16      setLEDs(B00000010, delayMS);
17      setLEDs(B00000001, delayMS);
18      // This is the end of the "cylon" pattern. Remember that loop() starts
            over after it's done, so the pattern repeats.
19    } // end of loop() - go back and start again
20
21    // This is the function that actually applies the pattern to the LED and
          then delays for the specified amount of time. This version takes a
          single byte as input that holds the ON/OFF values for the LEDs. It
          loops through each bit of that byte to set the LED values. The LED
          pin numbers are held in the global LEDPins array.
22    void setLEDs(byte LEDvalues, int delayMS) {
23      // You can access each bit of the LEDvalues byte using the Arduino
            syntax: bitRead(number, whichBit);
24      for(int i=0; i<8; i++) { // loop 8 times - i = 0,1,2,3,4,5,6,7
25        digitalWrite(ledPins[i], bitRead(LEDvalues, i));
26      }
27      delay(delayMS); // wait for delayMS milliseconds so you can see the
            change
28    } // end of setLEDs()
```

4.4.8 Dimming an LED with pulse width modulation

An interesting side note on LEDs is that they are not easily "dimmable." With an incandescent light bulb if you turn the voltage down, the bulb gets dimmer, and turning the voltage up results in a brighter light (until you go too far and the bulb burns out). This works for incandescent bulbs driven by either DC or AC power.

LEDs work differently. They start to conduct current when the voltage exceeds the Vf forward voltage, but changing the voltage has little effect on the light output because they are current-controlled devices (once the voltage is higher than Vf). There are minor variations in brightness as you increase the current, but mostly they are either on or off.

The good news about LEDs is that they turn on and off really fast. So, if you turn them on and off very quickly, they can look dimmer to our eyes. Our eyes are not as quick as an LED, so if they flash quickly, they look on, but dim because they are only on for half the time (for example). This technique is known a pulse width modulation or PWM. Using PWM the signal is pulsed on and off very quickly (on the order of 500Hz or 500 times per second for Arduino's PWM). The width of the pulse determines how much of the time the LED is on vs. off. This can simulate the effect of being at an intermediate voltage between 0v and 5v by adjusting the percentage of time that the PWM pulses are high and low.

Arduino has a function that controls this type of pulse width modulation called `analogWrite(pin, value)`. The pin argument is the digital pin to control. It turns out that there are some pins on the Arduino that `analogWrite` works with, and some that it does not. For this to work, you need to choose a pin that is a "PWM pin." On the Arduino Uno the PWM pins are 3, 5, 6, 9, 10, and 11 and are marked on the board with a ∼ in front of the pin number. The brightness value is an integer between 0 (fully off) and 255 (fully on). If the value is somewhere between 0 and 255, the PWM signal will be modulated to simulate a voltage between 0v and 5v by adjusting the pulses. When connected to an LED, this can provide a wide range of apparent brightness for that LED. Note that **analogWrite()** is ONLY usable on digital pins, not analog pins, and does not really produce an analog voltage, just a simulation of the analog voltage using PWM. Analog pins are used only as inputs on the Arduino.

Here is a snippet of code that fades the LED from full off to full on and back again using **analogWrite()**:

Listing 4.15 Fading LED

```
1  /* Use analogWrite() to fade an LED */
2  int ledPin = 10;          // LED on pin 10 (like Figure 4)
3  int fadeDelay = 30; // short delay to see the effect of the fading
4  void setup()  {
5    pinMode(ledPin, OUTPUT); // set pin 10 to be an output
6  }
7  void loop()  {
8    // fade LED from min to max in increments of 5 steps
9    for(int val=0; val<256; val+=5) {
10     analogWrite(ledPin, val); // sets the value (from 0 to 255)
11     delay(fadeDelay);         // wait a bit to see the effect
12   }
13   // fade LED from max to min in increments of 5 steps
14   for(int val=255; val>=0; val-=5) {
15     analogWrite(ledPin, val); // sets the value (from 0 to 255)
16     delay(fadeDelay);         // wait a bit to see the effect
17   }
18 }
```

4.4.9 Using randomness

I will include one final example that incorporates randomness into your LED lights. Using random numbers can add an element of spontaneity or unpredictability to your LED light flashing patterns. The Arduino IDE includes a function that returns random numbers: **random(min,max-1)**. The function we used before, **setLEDs(bits,delay)**, is a "side-effect" function. When you call it, it causes something to happen. The **random(min,max-1)** function is a "return-value" function. It doesn't cause anything to happen – instead it returns a different random number every time it is called. So, it can be used any place in your program you would normally have used a number, but in this case you will get a different number each time that code is executed.

The arguments to the function are the minimum value to be returned, and (somewhat confusingly) one more than the maximum number. That is, if you call **random(0,8)** you will get a random number between 0 and 7 returned when that function is executed. This mild confusion is because of the C-style array numbering that starts with zero. That is, an array with 8 elements has those elements numbered 0 through 7. So, you can pick a random element of the array starting at 0 and with 8 elements using the **random(0,8)** function for the index. You can argue with that choice, but that is the way the function was designed.

The last piece of code is a program that makes an LED flicker like a candle. To do this the **random(min,max-1)** function is used both to get random times to make the LED change brightness, and also to choose the brightness value randomly. Notice how the random function is used in place of a number, but allows a different, random, value to be used every time through the loop.

Listing 4.16 Candle flickering effect in LED

```
1  // A program to simulate a candle flickering effect on an LED.
2  // Use the random function to vary both the brightness and the time
3  // between changes to that brightness to simulate the random flickering
4  // of a candle.
5  int myLEDpin = 10; // define a global integer (int) variable to hold the
       pin number
6  void setup() {
7    pinMode(myLEDpin, OUTPUT);   // initialize digital pin 10 as an output.
8  }
9  void loop() {
10   // write a random value for the brightness using analogWrite to get
         different brightness
11   // and the random function to choose a different brightness value each
         time through the loop
12   analogWrite(myLEDpin, random(100, 256));  // random value from 100 to 255
13   delay(random(50, 151)); // delay for a random time from 50ms to 150ms
14 } // loop() will repeat forever
```

4.4.10 Summary – Questions you should ask about LEDs

Some questions that you should ask when connecting LEDs to a microcontroller like the Arduino are:

- What is the forward voltage (V_f) of the LED that you are trying to use?

  You can get this value from the vendor of the LEDs, or you can figure it out using a resistor, power supply, and voltmeter. To figure it out on your own, connect the LED through a largish resistor (470Ω for example) from +5v to ground. Measure the voltage drop across the lit LED. This is the forward voltage. A fancy multi-meter may also have a diode setting that directly measures the forward voltage. If you do not know and do not want to measure, you can usually safely assume roughly 2v as a normal forward voltage for a standard LED. Blue and white LEDs may be a little higher – as much as 3v or even 3.5v in some cases.

- How much current should you put through the LED?

  Your specification for the LED should tell you what the max current capability of your LED is. If it does not, assume that 20mA is the maximum. You can always back off a little bit from the LED's max without having much impact on the brightness. I usually size the current limiting resistor for 15-18mA.

- What current-limiting resistor should you use?

  You can compute the current-limiting resistor for a single LED using the formula

$$R_{current-limit} = (V_{source} - V_f)/I_{current-wanted}$$

  where V_{source} is the power supply voltage, V_f is the LED's forward voltage, and $I_{current-wanted}$ is the amount of current you would like to put through (or limit to) the LED. If your computation results in an odd-sized resistor that you do not have, choose the next larger resistor that you do have. It's always safer to make the resistor a little larger (limit the current a little more).

- How bright would you like your LED?

  Use `digitalWrite(pin, HIGH/LOW)` to set the LED full on or full off. Use `analogWrite(pin, value)` to set the LED to an intermediate brightness with value being between 0 and 255. This only works on Arduino pins 3, 5, 6, 9, 10, and 11 on the Uno boards (see Arduino.cc for details of other Arduino boards).

- How many LEDs can I turn on at once with an Arduino?

  The Arduino UNO is limited to 200mA total across all 14 digital pins. So, if you size your resistor for 20mA, you should only turn on 10 LEDs at the same time. If you size your resistor for 14mA then you could turn on all 14 at the same time safely ($14 \times 14 = 196$mA).

Similar projects

A link to a video showing a similar project created by Sean Flannery and Cody Gallager, where the LED Lights follow the name typed with a MORSE code can be found here: `https://vimeo.com/338101197`

CHAPTER 5

Translation

CONTENTS

5.1 CODE THE SAME THING IN MANY LANGUAGES

EACH coding language has its strengths and weaknesses, and platform dependencies. Or, put another way, no single language is good for all types of coding tasks or devices to run them on. Whilst C++ is a very robust language for building apps for Windows environments, it is not suitable for a Mac (you need to recompile in Xcode for it to run on a Mac). Python is great for quick prototyping and as a scripting tool inside other application environments (such as 3D modelling programs and web servers, and it runs in Windows, Mac, Linux). Java is similar to Python, but a more complete package. It will run on almost any device and integrates with other software such as web servers, network apps, and hardware controllers. Therefore, it is important to know what coding language to use for what task. Sometimes, the same application can be written in different languages with the same resulting output. In this section we provide an example of this – re-writing Fortran in Java before going back to Java for a thorough overview and explanation of how to handle images in Java.

5.2 TRANSLATION OF FORTRAN INTO JAVA

By Md Fahimul Islam
Walgreens, Illinois, USA

As discussed in the introduction, there are many coding languages available to learn. In practice you should not restrict yourself to learning just one language. Therefore, it might be a good learning experience to see how a code written in one language can be translated into another language. For this reason a segment of the code for a "school of fish" written by Anna Ursyn in Fortran was translated by Fahim Islam into Java resulting in the same output. This type of arrangement for coding graphics involved a coordinate system. The x, y, and zs were re-coded in Java. As each fish in the original code took a lot of numbers to describe, Fahim decided to translate only one fish. But the output looks the same as a fragment of the original graphic.

First, see the Fortran code along with the graphics in Figure 5.1.

and now the translation:

First, the selected fragment under discussion is shown in Figure 5.2.
Now, the code and an output follow:

Listing 5.1 Surface

```
1  import java.awt.Graphics;
2  import java.awt.Graphics2D;
3  import javax.swing.JFrame;
4  import javax.swing.JPanel;
5  import javax.swing.SwingUtilities;
6
7  class Surface extends JPanel {
8
9    private void doDrawing(Graphics g) {
10
11     double[] XM = { 11.3, 12.1, 12.6, 14.5, 15.2, 16.0, 17.0, 19.0, 20.2,
12         20.8, 21.4, 32.0, 21.4, 20.8, 21.5, 22.5, 23.8, 26.3, 28.5,
13         28.5, 29.5, 29.5, 30.5, 30.5, 31.3, 31.0, 32.0, 32.5, 32.0,
14         32.0, 34.6, 38.5, 37.0, 38.5, 37.5, 38.0, 37.5, 38.0, 37.0,
15         37.7, 38.0, 40.5, 39.7, 40.5, 39.0, 34.5, 32.0, 29.0, 27.0,
16         29.0, 32.0, 32.5, 31.3, 29.7, 29.0, 27.0, 23.5, 23.0, 22.5,
17         20.0, 22.5, 22.9, 22.4, 20.0, 17.0, 15.0, 14.0, 13.0, 12.0,
```

FIGURE 5.1 Fortran code and a graphical output. © Anna Ursyn. Used with permission.

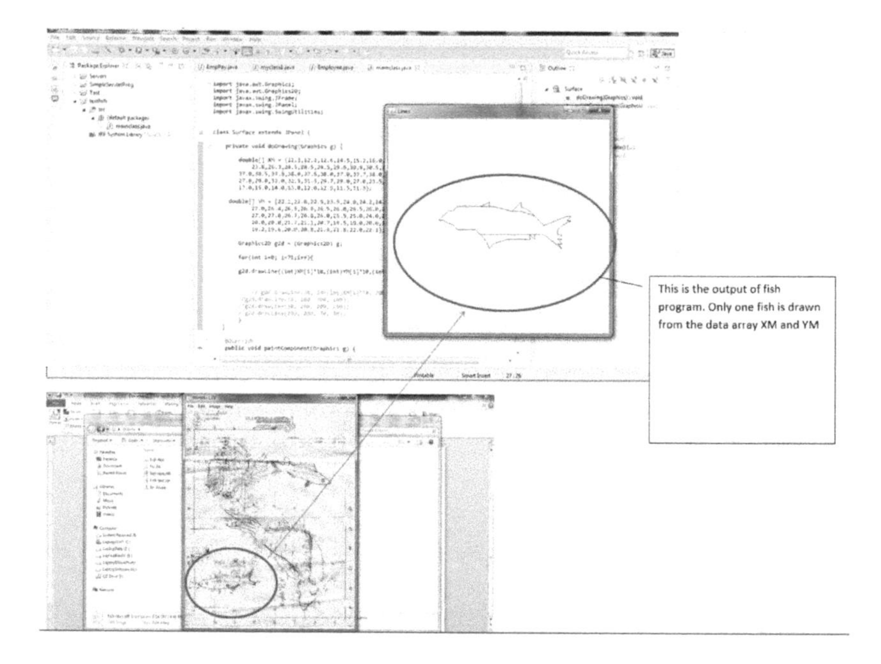

FIGURE 5.2 Fish program; an oval shows the singled out fish for the translation into Java. © Md Fahimul Islam. Used with permission.

```
18            12.5, 11.5, 11.3 };
19
20        double[] YM = { 22.1, 22.8, 22.9, 23.9, 24.0, 24.2, 24.1, 24.9, 25.1,
21            25.7, 25.0, 25.0, 25.0, 25.7, 26.0, 28.0, 27.0, 26.4, 26.5,
22            26.0, 26.5, 26.0, 26.5, 26.0, 26.1, 25.6, 26.0, 26.1, 25.3,
23            25.0, 25.0, 28.0, 27.0, 27.0, 26.7, 26.8, 26.0, 25.5, 25.0,
24            24.0, 23.5, 22.7, 22.0, 21.5, 21.7, 23.0, 21.7, 20.0, 20.0,
25            20.0, 21.7, 21.1, 20.7, 19.5, 18.0, 20.0, 19.1, 19.0, 19.0,
26            19.4, 19.0, 18.0, 18.0, 19.4, 19.2, 19.6, 20.0, 20.8, 21.4,
27            21.8, 22.0, 22.1 };
28
29        Graphics2D g2d = (Graphics2D) g;
30
31        for (int i = 0; i < 71; i++) {
32
33          g2d.drawLine((int) XM[i] * 10, (int) YM[i] * 10,
34              (int) XM[i + 1] * 10, (int) YM[i + 1] * 10);
35
36          // g2d.drawLine(30, 30+(int)XM[i]*10, 200, 30+i);
37          // g2d.drawLine(30, 100, 300, 100);
38          // g2d.drawLine(30, 200, 200, 200);
39          // g2d.drawLine(200, 200, 30, 30);
40        }
41      }
42
43      @Override
44      public void paintComponent(Graphics g) {
45
46        super.paintComponent(g);
47        doDrawing(g);
48      }
```

```
49  }
50
51  public class mainclass extends JFrame {
52
53    public mainclass() {
54
55      initUI();
56    }
57
58    private void initUI() {
59
60      setTitle("Lines");
61      setDefaultCloseOperation(JFrame.EXIT_ON_CLOSE);
62
63      add(new Surface());
64
65      setSize(500, 500);
66      setLocationRelativeTo(null);
67    }
68
69    public static void main(String[] args) {
70
71      SwingUtilities.invokeLater(new Runnable() {
72
73        @Override
74        public void run() {
75
76          mainclass lines = new mainclass();
77          lines.setVisible(true);
78        }
79      });
80    }
81  }
```

Explanation for the code is provided next. The purpose of this is to encourage those who are willing to learn Java programming to apply their own algorithm to manipulate an image. For beginners, this article includes step-by-step directions to start Java programming. Later it shows how Image processing can be accomplished in your own Java program. The example code shows how to store an image file in a two-dimensional-array, display it, convert an image to greyscale and save the new image to a new jpeg file which is part of the first two phases of Optical Character Recognition (OCR).

5.3 IMAGE PROCESSING IN JAVA FROM SCRATCH

Image processing is necessary in Pattern Recognition, OCR, Image Editing, Computer Vision and many other relevant areas. According to this project, Java image processing refers to the practice of image manipulation in the Java programming language. Programming is an art that can be learned following step-by-step instructions as for any programming language. The main objective of this chapter is aimed at illustrating the basic steps of Java programming with the ability to read, store, display, process, and save an image file in a Java program. Reading and Storing an image file also refers to image acquisition, which is the first phase of any image processing technique. In OCR the first two phases are image acquisition and greyscale conversion, which is what we will be doing in this project. This is similar to the conversion of a colour image to greyscale using Image Editing software (such as Adobe Photoshop).

TABLE 5.1 Some useful methods when working with images

Method	Description
`getHeight()`	Returns the height of the `BufferedImage`
`getWidth()`	Returns the width of the `BufferedImage`
`getRGB(int x, int y)`	Returns an integer pixel in the default RGB colour model
`setRGB(int x, int y, int rgb)`	Sets a pixel in this `BufferedImage` to the specified RGB value

5.3.1 Digital images in Java

A digital image is a set of values representing a two-dimensional array of pixels stored in a computer file. The set of values is organised according to rows and columns of image pixels. In Java, the `BufferedImage` class provides different methods and properties to access and manipulate image object data. In this chapter the following methods are used from the `BufferedImage` class (see Table 5.1).

5.3.2 Digital image processing

Digital image processing refers to the use of computer algorithms to perform image processing on digital images. As a subcategory or field of digital signal processing, digital image processing has many advantages over analogue image processing.

Digital image processing allows the use of much more complex algorithms, and hence, can offer both more sophisticated performance for simple tasks, and the implementation of methods which would be impossible by analogue means.

In particular, digital image processing is the only practical technology for:

- Classification
- Feature extraction
- Pattern recognition
- Projection
- Multi-scale signal analysis

5.3.3 Pattern recognition

Following from the above-mentioned practical technologies, pattern recognition is a branch of machine learning that focuses on the recognition of patterns and regularities in data, although it is in some cases considered to be nearly synonymous with machine learning. Pattern recognition systems are in many cases trained from labelled "training" data (supervised learning), but when no labelled data are available, other algorithms can be used to discover previously unknown patterns (unsupervised learning).

The terms pattern recognition, machine learning, data mining and knowledge discovery in databases (KDD) are hard to separate, as they largely overlap in their scope. Machine learning is the common term for supervised learning methods and originates from artificial intelligence, whereas KDD and data mining have a larger focus on unsupervised methods and stronger connection to business use. Pattern recognition has its origins in engineering, and the term is popular in the context of computer vision: A leading computer vision conference is named Conference on Computer Vision and Pattern Recognition. In pattern recognition, there may be a higher interest in formalising, explaining and visualising the pattern, whereas machine learning traditionally focuses on maximising the recognition rates. Yet, all of these domains have evolved substantially from their roots in artificial intelligence, engineering and statistics, and have become increasingly similar by integrating developments and ideas from each other.

5.3.4 Computer vision

Computer vision is the analysis of digital images so as to extract information automatically. The extracted information may be trivially simple, such as an answer to the question, "What colour is this?" or it may be much more complex, such as "Whose face is this?"

5.3.5 Optical character recognition

Optical character recognition (OCR) is the mechanical or electronic conversion of images of typed, handwritten or printed text into machine-encoded text. It is widely used as a form of data entry from printed paper data records, whether passport documents, invoices, bank statements, computerised receipts, business cards, mail, printouts of static-data, or any suitable documentation.

OCR software engines have been developed into many kinds of object-oriented OCR applications.

They can be used for:

- Data entry for business documents, e.g. check, passport, invoice, bank statement and receipt

- Automatic number plate recognition
- Automatic insurance documents key information extraction
- Extracting business card information into a contact list
- Making textual versions of printed documents more quickly, e.g. book scanning for Project Gutenberg
- Make electronic images of printed documents searchable, e.g. Google Books
- Converting handwriting in real time to control a computer (pen computing)
- Defeating CAPTCHA anti-bot systems, though these are specifically designed to prevent OCR
- Assistive technology for blind and visually impaired users

5.4 BASIC STEPS OF JAVA PROGRAM DEVELOPMENT

The following outlines how to get ready to do some Java programming. The steps required for developing any Java application are as follows:

1. Java compiler download and installation
2. An IDE Software download and installation where the code will be written (or you can use the CMD)
3. Writing code to develop application
4. Compile and run application

After downloading the Java compiler to run your program or application steps are described as follows:

5.4.1 Java compiler download and installation

To download the Java compiler, Google Java and download appropriate compiler from the Oracle site (see Figure 5.3). Download and install the Java SE Platform (JDK). There are several tutorials that can be followed depending on your operating systems.

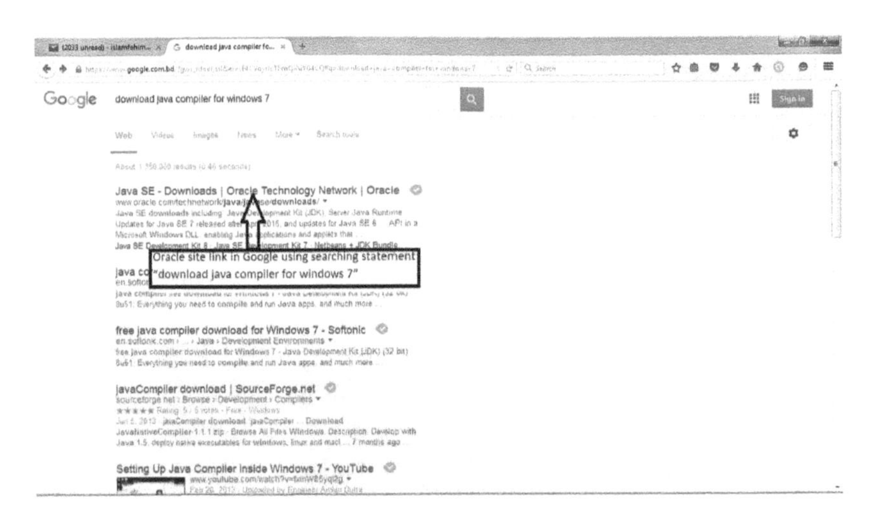

FIGURE 5.3 Search Java compiler link. © Md Fahimul Islam. Used with permission.

5.4.2 IDE software download and installation

An Integrated Development Environment(IDE) is the platform where the basic task of application development is performed, such as code writing, compiling, developed application running and debugging. In this project, Eclipse software is recommended as the Java IDE. You can download the Eclipse IDE for Java developing from `https://eclipse.org/downloads`. Make sure you select the appropriate link for 32bit or 64bit Operating Systems. The site provides instruction on the steps to install Eclipse (see Figure 5.4).

Before launching, the program will ask for a workspace folder where the necessary source code will be saved. Select or name a folder and click ok to start Eclipse IDE install. Close the welcome screen and it will show the IDE interface as follows:

Go to File > New > Click on Java Project. A new window will pop up to enter project name and other options. Enter the project name *JavaImageProc* and click finish, shown in Figure 5.5.

From the Package Explorer, click on the **`JavaImageProc`** right arrow to Explore src. Right click on src > new > click class. It will prompt for the name of the class file as in Figure 5.6.

Enter the class name as **`ImageProcessing`** as in the above picture and click finish. Now explore src, and under the (default package) > **`ImageProcessing.java`** will be created. Double clicking the file will open the source code file in the middle window (see Figure 5.7).

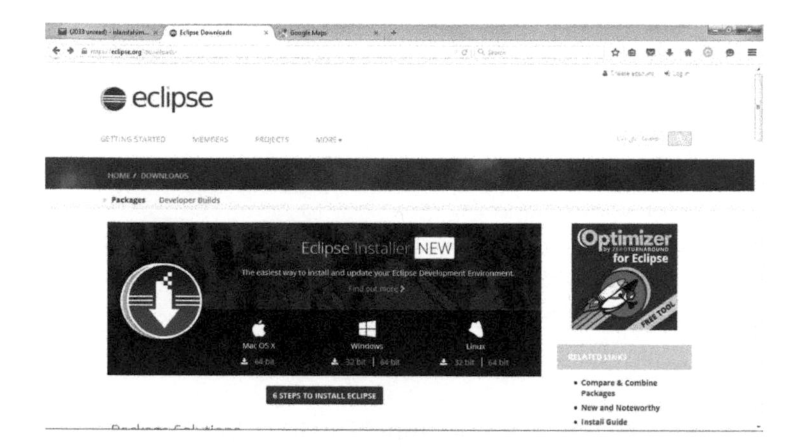

FIGURE 5.4 Eclipse IDE download. © Md Fahimul Islam. Used with permission.

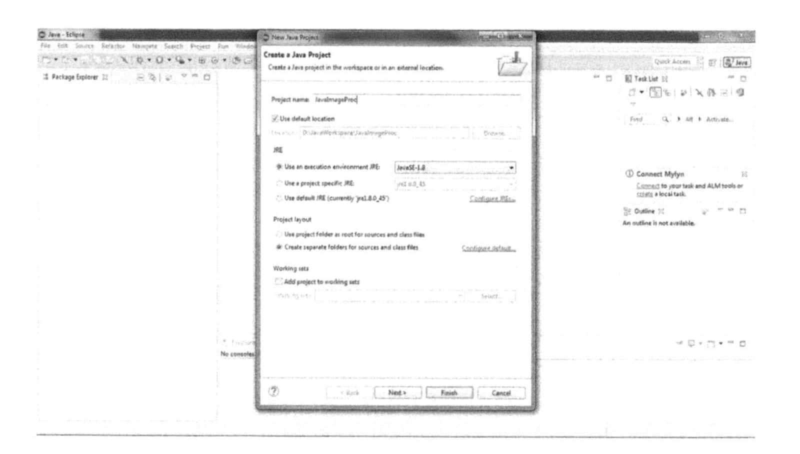

FIGURE 5.5 Create Java project. © Md Fahimul Islam. Used with permission.

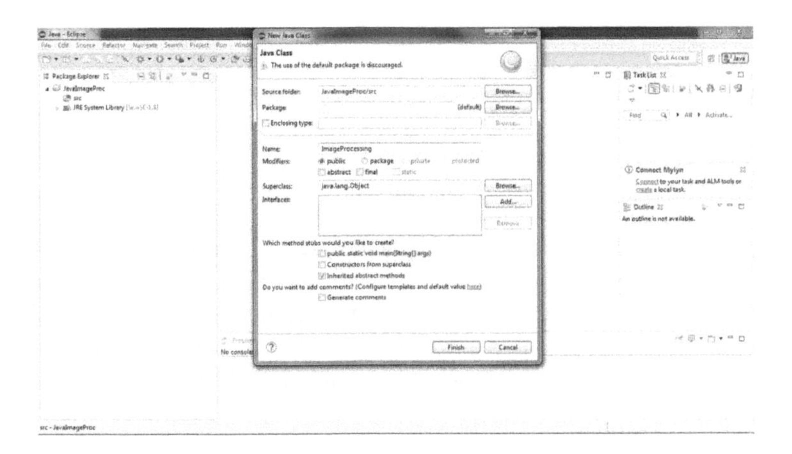

FIGURE 5.6 Add Java source file. © Md Fahimul Islam. Used with permission.

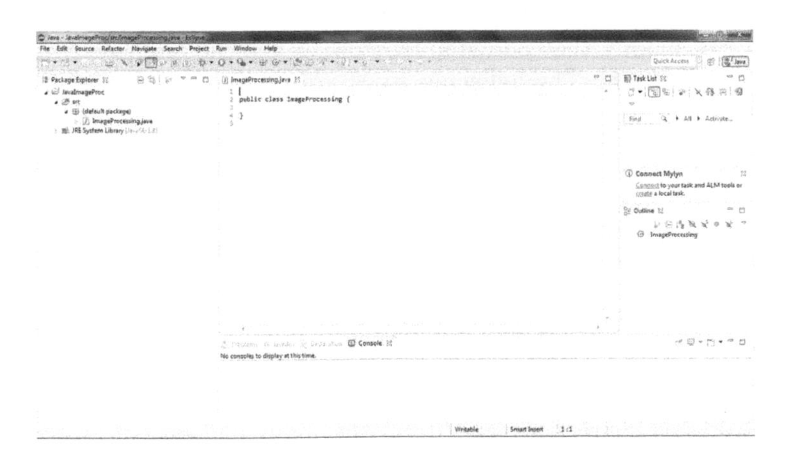

FIGURE 5.7 Open Java source file. © Md Fahimul Islam. Used with permission.

5.4.3 Writing code to develop application

Now start writing code by first including a main function as follows:

```
public class ImageProcessing {

  public static void main(String[] args) {
    System.out.println("Start image processing");
  }
}
```

The above code will print a line in the console. The above code has a public class named `ImageProcessing`. The class name and the file name should be the same as per the standard Java programming requirements. Then we define the main function. The class and main function are required for the Java programming structure, but we will only focus on the `println` statement, which is directly responsible for what the output will be.

5.4.4 Compile and run application

After the code has finished writing, press Ctrl+F11 to use any other method to run the application. Before compilation and running, another window will pop up for saving the source file. Press ok, and the output of the program will show in the console window as follows:

```
Start image processing
```

5.5 STEPS OF JAVA IMAGE PROCESSING

Now we can modify the code step-by-step to help us implement an image processing algorithm. The following steps are needed to organise the image processing steps in Java:

1. Draw a `JFrame` where the input and output image will be displayed.
2. Read an image file, draw the image in a `JPanel` and add to the `JFrame` to display graphically.
3. Store the image file to a two dimensional array and draw pixel by pixel into the `JPanel` to display graphically.
4. Modify the image array by applying an image processing algorithm (Greyscale conversion) and draw the input and output image to the `JPanel`.
5. Save the image file satisfying all the above steps.

The above steps produce the results illustrated next.

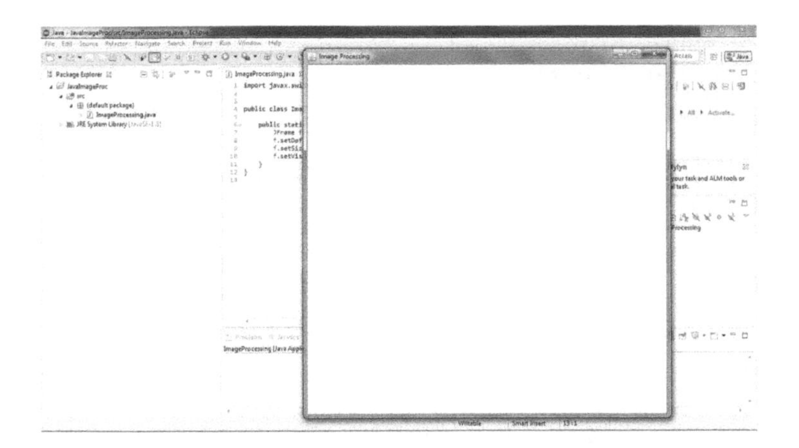

FIGURE 5.8 Draw a JFrame. © Md Fahimul Islam. Used with permission.

5.5.1 Draw a JFrame where the input and output image will be displayed

At first we need a frame to display the input and output image. For that, the following code creates a `JFrame` object with the title as `ImageProcessing`. After running the following code you should see a Graphical frame displayed. We can easily replace the previous source code with the following code to display a frame where we can further display an image file (see Figure 5.8).

```
1  import javax.swing.JFrame;
2
3  public class ImageProcessing {
4
5    public static void main(String[] args) {
6      JFrame f = new JFrame("Image Processing"); //Create f object
7      f.setDefaultCloseOperation(JFrame.EXIT_ON_CLOSE);//set default close
           operation
8      f.setSize(700,700);//set size of the frame
9      f.setVisible(true);//set f to visibility
10   }
11 }
```

5.5.2 Read an image file, draw the image in a JPanel and add to the JFrame to display graphically

Now we can extend the above code for several tasks to display an image by using `JPanel`. This is done by extending the previous `ImageProcessing` class to `JPanel`. From the following code, locate `public static void main(Strings [] args)` which is the header for the main function definition. The first line within this main function is the first line where the execution will be started as per Java methods (see Figure 5.9).

Listing 5.2 Image Processing #1

```
import javax.imageio.ImageIO;
import javax.swing.JFrame;
import javax.swing.JPanel;

import java.awt.Color;
import java.awt.Graphics;
import java.awt.image.BufferedImage;
import java.io.File;
import java.io.IOException;

public class ImageProcessing extends JPanel{

    BufferedImage img;

    ImageProcessing()
    {
        try{
            img = ImageIO.read(new File("d:\\input.jpg"));
        }
        catch (IOException e) {
            System.out.println(e.getMessage());
            }
    }


    public void paintComponent(Graphics g)
    {
    super.paintComponent(g);
    this.setSize(400,400);
    this.setBackground(Color.WHITE);

    g.drawImage(img, 10, 10, null);
    }



    public static void main(String[] args) {

        ImageProcessing IP = new ImageProcessing();

        JFrame f = new JFrame("Image Processing");
        f.setDefaultCloseOperation(JFrame.EXIT_ON_CLOSE);

        f.add(IP);

        f.setSize(700,700);
        f.setVisible(true);
    }
}
```

Therefore, `ImageProcessing IP = new ImageProcessing();` is the first line and it will create an object `IP`. It also invokes the constructor, and the execution control will be transferred to the below **`try-catch`** block to read **`input.jpg`** file from, say, the D drive (feel free to chance the location if you would like):

```
try{
  img = ImageIO.read(new File("d:\\input.jpg"));
} catch (IOException e) {
  System.out.println(e.getMessage());
}
```

FIGURE 5.9 Draw an Image to JPanel extending JFrame. © Md Fahimul Islam. Used with permission.

We can see in this portion of the code, the image file is read as **img** object. Now the execution control will be returned to the main function again and execute the following to create **JFrame**.

```
1  JFrame f = new JFrame("Image Processing");
2  f.setDefaultCloseOperation(JFrame.EXIT_ON_CLOSE);
```

Now **f.add(IP);** will add the **JPanel** class to the **JFrame**. Before adding the following method, it will be executed and the code row or line of code: **g.drawImage(img, 10, 10, null);** will draw the **img** object to the **JPanel**:

```
1  public void paintComponent(Graphics g)
2  {
3    super.paintComponent(g);
4    this.setSize(400,400);
5    this.setBackground(Color.WHITE);
6    g.drawImage(img, 10, 10, null);
7  }
```

5.5.3 Store the image file to a two dimensional array and draw pixel by pixel into the JPanel to display graphically

In this step, a two-dimensional array is used to hold the image data. The constructor is used to set the size of the two dimensional array for the image data.

```
1  height = img.getHeight();     //get the height value of the image
2  width = img.getWidth(); //get the width value of the image
3  image = new int[width][height]; //set the width-height dimension of the
       image array
```

The above code will be found in the constructor method which initialises the dimensions of the image array.

```
1 for(int y=0;y<height;y++)
2   for(int x=0;x<width;x++)
3     image[x][y]=img.getRGB(x,y);
```

The above code sets all the colour values to the image array. The `getRGB()` method returns RGB colour from the respective `x`, `y` coordinates. So by the above two sets of code we can transform the image into a two-dimensional array. Now it is time to draw the image on the `JPanel` from the image array. For that, we need to focus on the following code in the `paintComponent(Graphics g)` method:

```
1 for(int y=0;y<height;y++)
2   for(int x=0;x<width;x++)
3   {
4     g.setColor(new Color(image[x][y]));
5     g.drawLine(x+10, y+10, x+10, y+10);//draw image pixel by pixel
6
7   }
```

All together the following code might replace the previous step's code completely to get a new way of displaying the output (see Figure 5.10).

Listing 5.3 Image Processing #2

```
1  import javax.imageio.ImageIO;
2  import javax.swing.JFrame;
3  import javax.swing.JPanel;
4
5  import java.awt.Color;
6  import java.awt.Graphics;
7  import java.awt.image.BufferedImage;
8  import java.io.File;
9  import java.io.IOException;
10
11 public class ImageProcessing extends JPanel {
12
13   BufferedImage img;
14   int height;
15   int width;
16   int image[][];
17
18   ImageProcessing() // Constructor method
19   {
20     try {
21       img = ImageIO.read(new File("d:\\input.jpg"));
22       height = img.getHeight();
23       width = img.getWidth();
24       image = new int[width][height];
25
26       for (int y = 0; y < height; y++)
27         for (int x = 0; x < width; x++)
28           image[x][y] = img.getRGB(x, y);// storing the image data to
29       // a two-dimensional-array
30     } catch (IOException e) {
31       System.out.println(e.getMessage());
```

FIGURE 5.10 Draw Image pixel by pixel from two dimensional array. © Md Fahimul Islam. Used with permission.

```
32       }
33     }
34
35     public void paintComponent(Graphics g) {
36       super.paintComponent(g);
37       this.setSize(400, 400);
38       this.setBackground(Color.WHITE);
39
40       for (int y = 0; y < height; y++)
41         for (int x = 0; x < width; x++) {
42           g.setColor(new Color(image[x][y]));
43           g.drawLine(x + 10, y + 10, x + 10, y + 10); // draw image pixel
44                                   // by pixel
45
46         }
47     }
48
49     public static void main(String[] args) {
50
51       ImageProcessing IP = new ImageProcessing();
52
53       JFrame f = new JFrame("Image Processing");
54       f.setDefaultCloseOperation(JFrame.EXIT_ON_CLOSE);
55
56       f.add(IP);
57
58       f.setSize(700, 700);
59       f.setVisible(true);
60     }
61   }
```

5.5.4 Modify the image array by applying an image processing algorithm (greyscale conversion) and draw the input and output image to the JPanel

In this next step we can use a new method to apply our image processing algorithm. For that, we can add a new method **GrayScaleConvert()** and another array(**imgGrayScale[ ][ ]**) to store the changed data as follows:

```
1  public void GrayScaleConvert() // method for image processing (greyscale
       conversion)
2  {
3    int rd, gr, bl, grayLevel, grayPix;
4
5    for (int y = 0; y < height; y++)
6    for (int x = 0; x < width; x++) {
7      rd = (image[x][y] >> 16) & 0xFF;
8      gr = (image[x][y] >> 8) & 0xFF;
9      bl = (image[x][y] & 0xFF);
10
11     grayLevel = (rd + gr + bl) / 3;
12     grayPix = (grayLevel << 16) + (grayLevel << 8) + grayLevel;
13     imgGrayScale[x][y] = grayPix;
14
15   }
16 }
```

Basically, the above method says: For each pixel, hold the value of red, green and blue, and iterate by two in a for loop. Bitwise operation right shift (>>) and bitwise operation AND (&) are used to extract a colour value from each pixel. Each 24 bit integer holds three lots of colour information in each 8 bit at the binary level. We have to average the three colour values and supply the averaged value to each channel to assign **grayPix**, hence assign **imgGrayScale[x][y]** array. Then, if we print the **imgGrayScale[x][y]**, the grayscale image is obtained as follows:

```
1  for(int y=0;y<height;y++)
2    for(int x=0;x<width;x++)
3    {
4      g.setColor(new Color(imgGrayScale[x][y]));
5      g.drawLine(x+width+20, y+10, x+width+20, y+10);
6
7    }
```

Putting all together we can replace the previous step's code by the following (see output in Figure 5.11):

Listing 5.4 Image Processing #3

```
1  import javax.imageio.ImageIO;
2  import javax.swing.JFrame;
3  import javax.swing.JPanel;
4
5  import java.awt.Color;
6  import java.awt.Graphics;
```

```
7   import java.awt.image.BufferedImage;
8   import java.io.File;
9   import java.io.IOException;
10
11  public class ImageProcessing extends JPanel {
12
13    BufferedImage img;
14    int height;
15    int width;
16    int image[][];
17    int imgGrayScale[][];
18
19    ImageProcessing() {
20      try {
21        img = ImageIO.read(new File("d:\\input.jpg"));
22        height = img.getHeight();
23        width = img.getWidth();
24        image = new int[width][height];
25        imgGrayScale = new int[width][height];
26
27        for (int y = 0; y < height; y++)
28          for (int x = 0; x < width; x++)
29            image[x][y] = img.getRGB(x, y);
30
31      } catch (IOException e) {
32        System.out.println(e.getMessage());
33      }
34    }
35
36    public void GrayScaleConvert() // method for image processing (greyscale
37                   // conversion)
38    {
39      int rd, gr, bl, grayLevel, grayPix;
40
41      for (int y = 0; y < height; y++)
42        for (int x = 0; x < width; x++) {
43          rd = (image[x][y] >> 16) & 0xFF;
44          gr = (image[x][y] >> 8) & 0xFF;
45          bl = (image[x][y] & 0xFF);
46
47          grayLevel = (rd + gr + bl) / 3;
48          grayPix = (grayLevel << 16) + (grayLevel << 8) + grayLevel;
49          imgGrayScale[x][y] = grayPix;
50
51        }
52    }
53
54    public void paintComponent(Graphics g) {
55      super.paintComponent(g);
56      this.setSize(400, 400);
57      this.setBackground(Color.WHITE);
58      for (int y = 0; y < height; y++)
59        // Paint original photo
60        for (int x = 0; x < width; x++) {
61          g.setColor(new Color(image[x][y]));
62          g.drawLine(x + 10, y + 10, x + 10, y + 10);
63
64        }
65      for (int y = 0; y < height; y++)
66        // Paint grayscale photo
67        for (int x = 0; x < width; x++) {
68          g.setColor(new Color(imgGrayScale[x][y]));
69          g.drawLine(x + width + 20, y + 10, x + width + 20, y + 10);
70
71        }
72    }
73
```

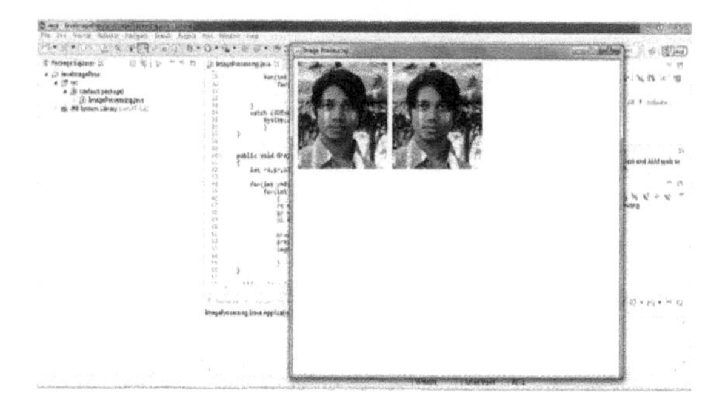

FIGURE 5.11 Modify the image, draw input and output image. © Md Fahimul Islam. Used with permission.

```
74      public static void main(String[] args) {
75
76        ImageProcessing IP = new ImageProcessing();
77
78        IP.GrayScaleConvert();// call image processing method
79
80        JFrame f = new JFrame("Image Processing");
81        f.setDefaultCloseOperation(JFrame.EXIT_ON_CLOSE);
82
83        f.add(IP);
84
85        f.setSize(700, 700);
86        f.setVisible(true);
87      }
88    }
```

5.5.5 Save the image file satisfying all the above steps

Saving the image is simple; we can set the greyscale data to `img` object and write a new file as per the following method:

```
1    public void OutputImage() // Output gray scale data to an image file
2    {
3      for (int y = 0; y < height; y++)
4      for (int x = 0; x < width; x++) {
5        img.setRGB(x, y, imgGrayScale[x][y]); // set GrayScale data to
6        // img object
7      }
8      try {
9        File o = new File("d:\\output.jpg");
10       ImageIO.write(img, "jpg", o); // Write img object to new image file
11     } catch (IOException e) {
12     System.out.println(e.getMessage());
13   }
```

All we have to do is call the above method from the main function and we obtain the following code that should replace all previous code to get the final result (see output in Figure 5.12):

Listing 5.5 Image Processing #4

```
1   import javax.imageio.ImageIO;
2   import javax.swing.JFrame;
3   import javax.swing.JPanel;
4
5   import java.awt.Color;
6   import java.awt.Graphics;
7   import java.awt.image.BufferedImage;
8   import java.io.File;
9   import java.io.IOException;
10
11  public class ImageProcessing extends JPanel {
12
13    BufferedImage img;
14    int height;
15    int width;
16    int image[][];
17    int imgGrayScale[][];
18
19    ImageProcessing() {
20      try {
21        img = ImageIO.read(new File("d:\\input.jpg"));
22        height = img.getHeight();
23        width = img.getWidth();
24        image = new int[width][height];
25        imgGrayScale = new int[width][height];
26
27        for (int y = 0; y < height; y++)
28          for (int x = 0; x < width; x++)
29            image[x][y] = img.getRGB(x, y);
30
31      } catch (IOException e) {
32        System.out.println(e.getMessage());
33      }
34    }
35
36    public void GrayScaleConvert() // method for gray scale conversion
37    {
38      int rd, gr, bl, grayLevel, grayPix;
39
40      for (int y = 0; y < height; y++)
41        for (int x = 0; x < width; x++) {
42          rd = (image[x][y] >> 16) & 0xFF;
43          gr = (image[x][y] >> 8) & 0xFF;
44          bl = (image[x][y] & 0xFF);
45
46          grayLevel = (rd + gr + bl) / 3;
47          grayPix = (grayLevel << 16) + (grayLevel << 8) + grayLevel;
48          imgGrayScale[x][y] = grayPix;
49
50        }
51    }
52
53    public void paintComponent(Graphics g) {
54      super.paintComponent(g);
55      this.setSize(400, 400);
56      this.setBackground(Color.WHITE);
57      for (int y = 0; y < height; y++)
58        for (int x = 0; x < width; x++) {
59          g.setColor(new Color(image[x][y]));
60          g.drawLine(x + 10, y + 10, x + 10, y + 10);
61
62        }
63      for (int y = 0; y < height; y++)
64        for (int x = 0; x < width; x++) {
65          g.setColor(new Color(imgGrayScale[x][y]));
66          g.drawLine(x + width + 20, y + 10, x + width + 20, y + 10);
```

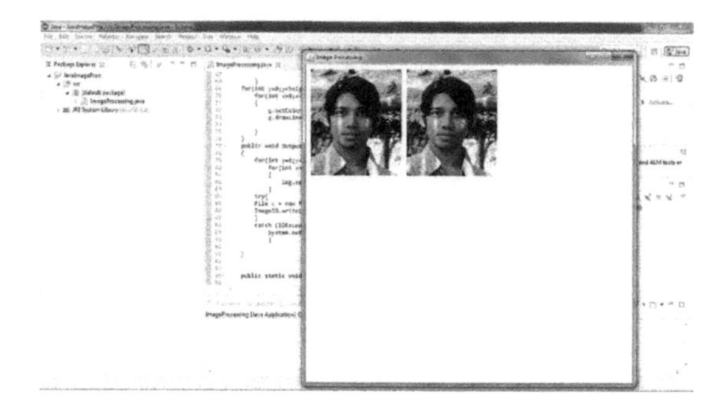

FIGURE 5.12 Save the modified image data to an output image file. © Md Fahimul Islam. Used with permission.

```
67
68          }
69      }
70
71      public void OutputImage() // Output gray scale data to an image file
72      {
73        for (int y = 0; y < height; y++)
74          for (int x = 0; x < width; x++) {
75            img.setRGB(x, y, imgGrayScale[x][y]);
76          }
77        try {
78          File o = new File("d:\\output.jpg");
79          ImageIO.write(img, "jpg", o);
80        } catch (IOException e) {
81          System.out.println(e.getMessage());
82        }
83
84      }
85
86      public static void main(String[] args) {
87
88        ImageProcessing IP = new ImageProcessing();
89
90        IP.GrayScaleConvert();
91        IP.OutputImage();
92
93        JFrame f = new JFrame("Image Processing");
94        f.setDefaultCloseOperation(JFrame.EXIT_ON_CLOSE);
95
96        f.add(IP);
97
98        f.setSize(700, 700);
99        f.setVisible(true);
100     }
101   }
```

Also find output.jpg file after image processing

5.5.6 Conclusion

A simple technique was used in this chapter to explore the image processing world. The above code might be used as the first step for image processing. A new algorithm might be your next step to practice OCR or pattern recognition. Changing the method `GrayScaleConvert()` could bring enormous data or information extraction from an image file. The resolution of the test image is 188 × 222 therefore a 400 × 400 `JPanel` was used to show the input and output image side by side. The size can be changed in the `paintComponent(Graphics g)` method for other sizes of image display.

5.5.7 Useful links

Below, you will find some relevant links which can be referred to while learning and improving your coding skills:

1. http://docs.oracle.com/javase/7/docs/api/java/awt/image/BufferedImage.html
2. https://docs.oracle.com/javase/tutorial/2d/images/index.html
3. https://en.wikipedia.org/wiki/Digital_image_processing
4. https://en.wikipedia.org/wiki/Pattern_recognition
5. https://en.wikipedia.org/wiki/Optical_character_recognition
6. http://stackoverflow.com/questions/9131678/convert-a-rgb-image-to-grayscale-image-reducing-the-memory-in-java
7. Jim R Parker. *Algorithms for Image Processing and Computer Vision.* John Wiley & Sons, 2010

CHAPTER 6

Conclusions and Educational Propositions

CONTENTS

THE aim of this book is to give the reader an insight into coding for artistic endeavours. It provides a diversity of examples with accompanying explanations. At one end of the spectrum, this book is pitched at those creative types who can see the benefits of learning to code but are too afraid to try it. At the other end, it also provides advanced projects that hopefully stretch the reader to try something more difficult than they would normally attempt by showing how easy it can be. Importantly, each project includes some visible output. We feel it is important to provide a tangible result rather than simply labouring over the principles of computer programming.

It is the current cohort of students and artists that are perhaps most affected by the proliferation of computer applications possible nowadays. That is because most of the non-IT demographic have had little to no formal training in computer programming. But, clearly, the possibilities for creative endeavours using computer programming are endless. Therefore, it is little wonder that they have been trying to teach themselves. It is hoped this book will provide some inspiration and guidance on the next steps. In the meantime, we also hope that some computer programming will be introduced very early into the curriculum to provide the next generation of artists with the skills they need to prosper in the world of digital media.

6.1 CODING IN CURRICULUM – GRAPHICAL THINKING

Some of the concepts and skills discussed in this book are important factors in constructing new school curricula. Learning visually with the use of coding and visualisation techniques supports the development of creativity and abstract thinking. Information can be presented in many ways: Numerical, graphical, or diagrammatic, as a sketch, drawing, diagram, plan, outline, image, geometric relationship, map, music and dance notation, object, interactive installation, or story. Visual metaphors, which make these concepts visible, are inherent in our thought, and thus enable visualisation of abstract concepts. Auditory metaphors are essential for music theory and appreciation. Abstract thinking usually refers to thinking about concepts and general principles that are separate from specific, concrete things. Abstract thinking supports the learning process, which involves creating representations of a new concept or re-presentations of existing knowledge.

While learning to write programs, one may use metaphors as well as analogies with one's everyday activities. Both verbal and visual analogies can be helpful in learning a code. For example, we looked at actions such as 'get', 'put', and connections such as 'while', 'if', or 'then'. To better understand how these metaphors and analogies for instruction are used in programming, we could divide a page into two columns with a specific issue in coding on the left side and a tangible analogy relating to things perceived through the senses on the right side. As an example, in the chapter "Metaphors for Dance and Programming: Rules, Restrictions, and Conditions for Learning and Visual Outcomes" the authors discuss some restrictive rules and can be compared to several approaches to coding [50]. For instance, some curricula promote parallel coding, where two students are paired up to learn together.

There are many solutions related to knowledge visualisations. For example tree-like structures used to show interconnections [3, 31] and multidimensional visualisations such as Minard's depiction of Napoleon's campaign [28] showing six variables on one graph, allowing one to write a book on the subject. It serves as visualisation tools, which is also used for eye tracking. They can be used to promote student interest and understanding of scientific concepts and processes. Scientists, such as Kepler, had to transform their visions into physical or mathematical solutions. Studying calculus (especially differentiation and integration) is important for understanding many areas of science. For example, the speed of the heart's muscle cell activation and its resulting contraction are crucial for the heart muscle's efficiency in pumping blood. The slope of a graph showing such activation can be measured as the derivatives and integrals of this function, and the area of the region bounded by this graph can be calculated as an integral of this function. Thus, we can say derivatives and integers help to explain basic processes and events in our surroundings, such as a heartbeat or a car engine's acceleration. In this manner, students gain a tool that is conducive to them learning logical ways of comprehending new concepts rather than just memorising by rote. Theoretical research applying mathematics helps convert the unseen, unfelt, or unheard into the visible, palpable, or sending of sounds. Hence, we can go from the physical to digital or vice versa. In fact, we can examine these capacities and use this information not just for theoretical, practical, and computational solutions within the domains of physiology, neuroscience, cognitive science, cognitive psychology, sociology, anthropology, medicine, computer science, but also human perception, philosophy, and art.

From this, we can see there is a common way of thinking in various computer languages. Acquiring programming skills in various computer languages supports communication. Codes of different kinds are used and have always been used in many areas of activity, including music and language – with its many kinds of alphabet, grammar, and double articulation, or the book and font design. Calligraphy plays different perceptual, aesthetic, and psychological roles in particular languages. Eye tracking experiments confirm the importance of the text's shape, and how it affects our speed of reading.

Because of these insights we can make educational propositions regarding the future shaping of learning and instruction in our schools. First, learning algebra should start at the early elementary school level, and then acquiring programming skills should follow, in various computer languages, which could be linked with traditional tools such as the now old but familiar Lego Logo. Learning visually with the use of visualisation techniques should support visualising processes or products through drawing with colours and sketching. Serious video games should be designed for more purposes than entertainment or fitness. Chess playing should be encouraged. In general, the world is transitioning to a new phase in human history, the digital era. Students need to get ready to engage in this new era. The best way to do this is by learning how to code. You do not have to be a wiz at coding; you just need to know enough to understand what is going on in a virtual machine (a computer). If you know at least this much, you can get others to do the code hacking for you. But, to earn their respect in the first place, you need to know some coding yourself. Hopefully this book has given you that core knowledge to go into the new world with open eyes to the possibilities. On the other hand, if you find you love coding, then there is much left to be done. And hopefully, for you, this book has sparked your interest sufficiently to make you want to know more. There are plenty of resources that you can find online to take the next steps. Most importantly, coding can be creative. We hope through this book we have stimulated your senses to use coding for your own creative pursuits!

6.2 RESOURCES

- Code.org
- Codeacademy.com
- Sounds.org
- Royalty free music

  Bensound.com
- includes 3d scanned objects

  textures.com
- Giving and taking images

  cgtextures.com
- Processing.org
- How data can be presented

  https://informationisbeautiful.net
- Gapminder.org
- http://www.storytellingwithdata.com/book
- Great example of interactive learning environment

  https://www.ptable.com
- Brilliant

  https://brilliant.org/courses/computer-science-fundamentals/intro-to-algorithms
- WickEditor: This software is similar to Adobe Flash/Animate, which are developed for 2D web animation. This software is developed for creating animations, games, and anything in between.

  http://wickeditor.com
- OpenToonz is an animation program allowing the user to modify the source code freely.

  https://opentoonz.github.io/e/index.html
- APL is an array-oriented programming language with a concise syntax which allows the creation of programs with smaller lines of code.

  https://tryapl.org

The video in the following link shows Conway's Game Of Life in few lines of code. `https://youtu.be/a9xAKttWgP4`

If interested in Game of Life, the following provides a good introduction: `https://fiftyexamples.readthedocs.io/en/latest/life.html`

- Godot. It has the distinct advantages of being free with a no-strings-attached license. The high-brow reference to Samuel Beckett's play suggests that it is for artists and not just gamers.

  `https://godotengine.org`

- Mathematical Art Galleries (Bridges and JMM = Joint Mathematics Meeting)

  `http://gallery.bridgesmathart.org`

- ACM SIGGRAPH - International Conference on Computer Graphics and Interactive Techniques

  `https://www.siggraph.org`

- Eurographics. Eurographics is a Europe-wide professional computer graphics association. The association supports its members in advancing the state of the art in computer graphics and related fields such as multimedia, scientific visualisation and human-computer interaction.

  `https://www.eg.org`

- Ars Electronica. The Ars Electronica Centre is a centre for electronic arts run by Ars Electronica situated in Linz, Austria, at the northern side of the Danube opposite the city hall of Linz.

  `https://ars.electronica.art/news/en`

- The Leonardo Journal. Leonardo is a peer-reviewed academic journal published by the MIT Press covering the application of contemporary science and technology to the arts and music

  `https://www.leonardo.info`

6.3 FURTHER READING

In this section, you will find a list of books for further reading in different categories, on art and cognition as well as coding for artists' resources.

6.3.1 Books on art and cognition

The list below presents a number of books on the interlinked domains of art and cognition:

Josef Albers. *Interaction of Color*. Yale University Press, 2013.

Rudolf Arnheim. *Visual Thinking*. Univ. of California Press, 1969.

Rudolf Arnheim. *Art and the Visual Perception: A Psychology of the Creative Eye*. University of California Press, 1974.

Rudolf Arnheim. *The Power of the Center: A Study of Composition in the Visual Arts*. Univ. of California Press, 1983.

Jorge Luis Borges. *The Book of Imaginary Beings*. Random House, 2002.

Antonio Damasio. *Self Comes to Mind: Constructing the Conscious Brain*. Vintage, 2012.

Antonio Damasio and Gil B Carvalho. The nature of feelings: Evolutionary and neurobiological origins. *Nature Reviews Neuroscience*, 14(2):143, 2013.

Howard Gardner. *Frames of Mind: The Theory of Multiple Intelligences*. NY: Basics, 1983.

Howard E Gardner. *Multiple Intelligences: New Horizons in Theory and Practice*. Basic Books, 2008.

Howard Gardner and Gardner E. Art, mind, and brain: A cognitive approach to creativity, 1984. Basic Books, 2008

Alan A Stone. Extraordinary minds: Portraits of exceptional individuals and an examination of our extraordinariness. *American Journal of Psychiatry*, 156(2):328–329, 1999.

Steve Garner. *Writing on Drawing: Essays on Drawing Practice and Research*. Intellect Books, 2012.

Edvard I Moser, May-Britt Moser, and Yasser Roudi. Network mechanisms of grid cells. *Philosophical Transactions of the Royal Society B: Biological Sciences*, 369(1635):20120511, 2014.

Robert J Sternberg and Scott Barry Kaufman. *The Cambridge Handbook of Intelligence*. Cambridge University Press, 2011.

Edward R Tufte, Nora Hillman Goeler, and Richard Benson. *Envisioning Information*, volume 126. Graphics Press, Cheshire, CT, 1990

Edward R Tufte, and Stan Rifkin. Visual explanations: Images and quantities, evidence and narrative. *ISIS-International Review*, Philadelphia, 88(4):748–748, 1997

James Faure Walker. *Painting the Digital River: How an Artist Learned to Love the Computer.* Prentice Hall Professional, 2006

Semir Zeki. *Splendors and Miseries of the Brain: Love, Creativity, and the Quest for Human Happiness.* John Wiley & Sons, 2011

6.3.2 Books on coding for artists

In order to expand your understanding on areas that could enrich artists' artworks, the following is a good starting point:

John M Blain. *The Complete Guide to Blender Graphics: Computer Modeling and Animation.* AK Peters/CRC Press, 2014.

Margaret A Boden. *AI: Its Nature and Future.* Oxford University Press, 2016.

Stefan Bertschi, Sabrina Bresciani, Tom Crawford, Randy Goebel, Wolfgang Kienreich, Martin Lindner, Vedran Sabol, and Andrew Vande Moere. What is knowledge visualization? Perspectives on an emerging discipline. In *Information Visualisation (IV), 2011 15th International Conference on*, pages 329–336. IEEE, 2011.

Margaret A Boden. Skills and the appreciation of computer art. *Connection Science*, 28(2):131–138, 2016.

Michele Bousquet. Light and Color. In *Physics for Animators*, pages 133–178. CRC Press, 2015.

John Boxall. *Arduino Workshop: A Hands-On Introduction with 65 Projects.* No Starch Press, 2013

Derek Breen. *Creating Digital Animations: Animate Stories with Scratch!* John Wiley & Sons, 2016.

Jason R Briggs. *Python for Kids: A Playful Introduction to Programming.* No Starch Press, 2013.

Brian Christian and Tom Griffiths. *Algorithms to Live By: The Computer Science of Human Decisions.* Macmillan, 2016.

Roberto Dillon. *2D to VR with Unity5 and Google Cardboard.* AK Peters/CRC Press, 2017.

Flint Dille and John Zuur Platten. *The Ultimate Guide to Video Game Writing and Design.* Lone Eagle Publishing Company, 2007.

Martin Erwig. *Once Upon an Algorithm: How Stories Explain Computing.* MIT Press, 2017.

Ira Greenberg, Dianna Xu, and Deepak Kumar. *Processing: Creative Coding and Generative Art in Processing 2.* Apress, 2012.

Viktoria Greanya. *Bioinspired Photonics: Optical Structures and Systems Inspired by Nature.* CRC Press, 2015.

Robin Hanson. *The Age of Em: Work, Love, and Life when Robots Rule the Earth.* Oxford University Press, 2016.

Andy Harris. *Game Programming: The L Line, The Express Line to Learning.* Wiley, 2007.

Hanna Brady and Jarryd Huntley. *Game Programming for Artists.* A K Peters/CRC Press, 2017.

Clifford A Pickover. *Fractal 3D Magic.* Sterling, 2014. `http://www.geome-tree.com/`.

Yasmin B Kafai and Quinn Burke. *Connected Code: Why Children Need to Learn Programming.* MIT Press, 2014.

Ajay Kapur, Perry R Cook, Spencer Salazar, and Ge Wang. *Programming for Musicians and Digital Artists: Creating Music with ChucK.* Manning Publications Co., 2015.

Manuel Lima. *The Book of Trees: Visualizing Branches of Knowledge.* Princeton Architectural Press, 2014

Nick Montfort, Patsy Baudoin, John Bell, Ian Bogost, Jeremy Douglass, Mark C Marino, Michael Mateas, Casey Reas, Mark Sample, and Noah Vawter. *10 PRINT CHR $(205.5+ RND (1));: GOTO 10.* MIT Press, 2012.

Nick Montfort. *Exploratory Programming for the Arts and Humanities.* MIT Press, 2016.

Terrence Masson. *CG 101: A Computer Graphics Industry Reference.* New Riders Publishing, 1999.

Chris Pine. *Learn to Program.* Pragmatic Bookshelf, 2009.

Chris Solarski. *Interactive Stories and Video Game Art: A Storytelling Framework for Game Design.* AK Peters/CRC Press, 2017.

Casey Reas and Ben Fry. *Processing: A Programming Handbook for Visual Designers and Artists.* Number 6812. MIT Press, 2007.

Alan Thorn. *How to Cheat in Blender 2.7 x.* CRC Press, 2017.

Ray Toal, Rachel Rivera, Alexander Schneider, Choe Eileen, Ray Toal, Rachel Rivera, Alexander Schneider, Choe Eileen, Ray Toal, Rachel Rivera, et al. Execution in the kingdom of nouns. In *Programming Language Explorations*, volume 48, pages xiii–xvi. Unicode Consortium Greenwich, CT, USA, 2017.

Laurens Valk. *Lego Mindstorms Ev3 Discovery Book: A Beginner's Guide to Building and Programming Robots.* No Starch Press, 2014.

Bruce Wands. *Art of the Digital Age.* Thames & Hudson, 2007

Bruce Wands. *Digital Creativity: Techniques for Digital Media and the Internet.* John Wiley & Sons, 2002.

Jon Woodcock. *Coding Projects in Scratch.* DK Publishing (Dorling Kindersley), 2016.

References

[1] Edwin A Abbott. *Flatland: A Romance of Many Dimensions.* OUP Oxford, 2006.

[2] Mohammad Majid al Rifaie and Tim Blackwell. Reduced projection angles for binary tomography with particle aggregation. *Evolutionary Intelligence*, 9(3):67–79, 2016.

[3] Benjamin Bederson and Ben Shneiderman. *The Craft of Information Visualization: Readings and Reflections.* Morgan Kaufmann, 2003.

[4] Margaret A. Boden. *Mind as Machine: A History of Cognitive Science.* Oxford University Press, 2008.

[5] Michele Bousquet. *Physics for Animators.* CRC Press, 2015.

[6] John Boxall. *Arduino Workshop: A Hands-On Introduction with 65 Projects.* No Starch Press, 2013.

[7] Klaus Bredl, Amrei Groß, Julia Hünniger, and Jane Fleischer. The avatar as a knowledge worker? How immersive 3d virtual environments may foster knowledge acquisition. *Electronic Journal of Knowledge Management*, 10(1):15, 2012.

[8] Nick Carlson. *Image versus Text.* Unpublished class project, University of Northern Colorado, 2014.

[9] L. Cooke. *3D printed house in China can withstand an 8.0 earthquake*, June 28, 2016 (Accessed October 17, 2018). `http://inhabitat.com/3d-printed-house-in-china-can-withstand-an-8-0-earthquake/`.

[10] J. Copeland. *Alan Turing: the codebreaker who saved 'millions of lives.' BBC News Technology, 19 June, 2012.*, Accessed October 17, 2018. `http://www.bbc.co.uk/news/technology-18419691`.

[11] John Counsell. Pointing the finger; a role for hybrid representations in vr and video? In *Information Visualization, 2003. IV 2003. Proceedings. Seventh International Conference on*, pages 633–638. IEEE, 2003.

[12] Merriam-Webster Dictionary and Thesaurus. *An Encyclopedia Britannica Company*, Accessed October 17, 2018. `http://www.merriam-webster.com/`.

[13] Alireza Ebrahimi. *C++ Programming: Easy Ways*. American Press, 2003.

[14] James Faure Walker. *Painting the Digital River: How an Artist Learned to Love the Computer*. Prentice Hall Professional, 2006.

[15] Wolfgang K Giloi. Konrad zuse's plankalkuel: The first high-level, "non von neumann" programming language. *IEEE Annals of the History of Computing*, 19(2):17–24, 1997.

[16] Ernest Goldstein. *Understanding and Creating Art*. Garrard Pub. Co., Textbook Division, 1986.

[17] Maitland Graves. The Art of Color and Design. *Journal of Aesthetics and Art Criticism*, 2(5):66–68, 1942.

[18] Nenad Grujović, Milan Radović, Vladimir Kanjevac, Jelena Borota, G Grujović, and Dejan Divac. 3d printing technology in education environment. In *34th International Conference on Production Engineering*, pages 29–30, 2011.

[19] Harry Henderson. *Encyclopedia of Computer Science and Technology*. Infobase Publishing, 2009.

[20] Matthew D Hilchey and Raymond M Klein. Are there bilingual advantages on nonlinguistic interference tasks? Implications for the plasticity of executive control processes. *Psychonomic bulletin & review*, 18(4):625–658, 2011.

[21] John L Irwin, Joshua M Pearce, and Gerald Anzalone. Evaluation of reprap 3d printer workshops in k-12 stem. In *2015 ASEE Annual Conference & Exposition*, pages 26–696, 2015.

[22] Walter Isaacson. The real leadership lessons of Steve Jobs. *Harvard Business Review*, 90(4):92–102, 2012.

[23] Stephen Johnson. *Stephen Johnson on Digital Photography*. O'Reilly Media, Inc., 2006.

[24] Owen Jones. *The Grammar of Ornament*. B. Quaritch, 1868.

[25] Norton Juster. *The Dot and the Line: A Romance in Lower Mathematics*. Chronicle Books, 2001.

[26] Wassily Kandinsky. *Concerning the Spiritual in Art*. Courier Corporation, 2012.

[27] Donald E Knuth and Luis Trabb Pardo. "The Early Development of Programming Languages." In *A History of Computing in the Twentieth Century*, pages 197–273. Elsevier, 1980.

[28] Menno-Jan Kraak. *Mapping Time: Illustrated by Minard's Map of Napoleon's Russian Campaign of 1812*. 2014.

[29] Gérard A Langlet. Building the APL atlas of natural shapes. In *ACM SIGAPL APL Quote Quad*, volume 24, pages 134–147. ACM, 1993.

[30] S. Lavington. *Alan Turing: is he really the father of computing? BBC News Technology, 19 June, 2012.*, Accessed October 17, 2018. `http://www.bbc.co.uk/news/technology-18327261`.

[31] Manuel Lima. *The Book of Trees: Visualizing Branches of Knowledge*. Princeton Architectural Press, 2014.

[32] Lott-Lavigna. *Watch this giant 3D printer build a house. Wired*, September 21, 2015 (Accessed October 17, 2018). `http://www.wired.co.uk/article/giant-3d-printer-builds-houses`.

[33] Joshua Marinacci and Chris Adamson. *Swing Hacks: Tips and Tools for Killer GUIs*. O'Reilly Media, Inc., 2005.

[34] Rebecca Mercuri and Kevin Meredith. An educational venture into 3d printing. In *Integrated STEM Education Conference (ISEC), 2014 IEEE*, pages 1–6. IEEE, 2014.

[35] John S Murnane. The psychology of computer languages for introductory programming courses. *New Ideas in Psychology*, 11(2):213–228, 1993.

[36] N. Negroponte. *XO Project*, Accessed October 17, 2018. `https://en.wikipedia.org/wiki/OLPC_XO`.

[37] Jim R Parker. *Algorithms for Image Processing and Computer Vision*. John Wiley & Sons, 2010.

[38] R. Pease. *Alan Turing: Inquest's suicide verdict 'not supportable.' BBC News Technology, 19 June, 2012.*, Accessed October 17, 2018. `http://www.bbc.co.uk/news/science-environment-18561092`.

[39] S. Pruitt. *Ancient Greeks May Have Used World's First Computer To Predict the Future*, Accessed October 17, 2018. `https://www.history.com/news/ancient-greeks-may-have-used-worlds-first-computer-to-predict-the-future`.

[40] Bates Ramirez. *This 3D printed house goes up in a day for under $10,000. Singularity Hub*, 2018, March 18.

[41] R Ravikumar and I Mohsin Khan. Design & development of a 3d printer. In *Proceedings of 12th IRF International Conference*, 2015.

[42] Edward Regis. *Who Got Einstein's Office? Eccentricity and Genius at the Institute for Advanced Study.* Perseus Publishing, 1987.

[43] Dan Roam. *The back of the napkin: Solving problems and selling ideas with pictures.* Portfolio, 2013.

[44] Geoffrey Sampson. Schools of linguistics. *Linguistics*, 410(9):P61, 1980.

[45] Chelsea Schelly, Gerald Anzalone, Bas Wijnen, and Joshua M Pearce. Open-source 3-d printing technologies for education: Bringing additive manufacturing to the classroom. *Journal of Visual Languages & Computing*, 28:226–237, 2015.

[46] Herbert A Simon and Allen Newell. Human problem solving: The state of the theory in 1970. *American Psychologist*, 26(2):145, 1971.

[47] Howard Gardner and Gardner E. Art, mind, and brain: A cognitive approach to creativity, 1984. Basic Books, 2008.

[48] Edward Tufte. The visual display of quantitative information. *Graphics Press, Cheshire, USA,*, 4(5):6, 2001.

[49] Edward R Tufte, Nora Hillman Goeler, and Richard Benson. *Envisioning Information*, volume 126. Graphics Press, Cheshire, CT, 1990.

[50] Anna Ursyn, Mohammad Majid al Rifaie, and Md Fahimul Islam. Metaphors for dance and programming: rules, restrictions, and conditions for learning and visual outcomes. In *Knowledge Visualization and Visual Literacy in Science Education*, pages 255–305. IGI Global, 2016.

[51] Bruce Wands. *Art of the Digital Age.* Thames & Hudson, 2007.

[52] Richard L Wexelblat. *History of Programming Languages.* Academic Press, 2014.

[53] Alex W White. *The Elements of Graphic Design: Space, Unity, Page Architecture, and Type.* Skyhorse Publishing, Inc., 2011.

[54] Stephen Wolfram. *A New Kind of Science*, volume 5. Wolfram Media. Champaign, IL, 2002.

[55] Hannah Wright. Computer programmers and the "bilingual advantage": Enhanced executive control in non-linguistic interference tasks. *Institute of Education, University of London, London, UK*, 2012.

[56] Semir Zeki. Art and the brain. *Journal of Consciousness Studies*, 6(6-7):76–96, 1999.

[57] Semir Zeki. Artistic creativity and the brain. *Science*, 293(5527):51–52, 2001.

[58] Semir Zeki. *Splendors and Miseries of the Brain: Love, Creativity, and the Quest for Human Happiness*. John Wiley & Sons, 2011.

Glossary

Amplitude: how much a value changes over time, such as the volume of a sound

Analogue: physically manifest rather than virtually (a mechanical light switch rather than a digital button on a screen which needs to be pressed with a mouse cursor)

API: application programming interface

ASCII: American Standard Code for Information Interchange (basic text)

Augmented reality: extended reality, such as a glasses or goggles that overlay or superimpose one's view of the world with computer-generated images of other objects

Binary: has only two parts (zeros and ones, plus or minus, yes or no, etc.)

Bitmap: bit values arrayed such that when displayed on a monitor they generate a coloured image

Bits: smallest possible unit of information in a computer, represented by a zero or a one

Boolean: subtraction, addition, union (take one thing away from another, add one thing to another, make one thing out of two things)

Breadboard: usually a perforated fibreglass or other composite plastic board with no copper backing so various electronic components can be inserted and connected from underneath

Buffer: temporary storage of information in dynamic memory

Bytecode: the code that is processed by a program, such as java, C++, ADA

Bytes: a collection of binary digits or bits forming a unit of memory

CAD/CAM: Computer Aided Design/Computer-Aided Manufacturing

Cartesian coordinates: the three-dimensional coordinate system devised by Rene Descartes in the 17th century

Chiaroscuro: drawing technique developed in Renaissance to demonstrate shadows and shading in perspective image

Chip IC: integrated circuit

Circuitboard: usually fibreglass or other composite plastic board with copper tracks used as electrical conductors for various electronic components

Class: a programming object that provides instructions for how to operate on other objects

CMYK: cyan, magenta, yellow and black, colour gamut used in printing

Cognitive function: the mental processes used to make us aware of the world and make sense of our place in it

Colour ramp: changing of colour over distance

Compile: to convert human readable code into machine readable code

Contrasting or complementary colours: colours opposite each other on the RGB colour wheel

CPU: central processing unit

Curricular: the subjects or courses of study available at a learning institution (school, university, industry)

Dynamic memory: temporary memory (RAM)

Fibonacci number sequence: each subsequent number is the addition of the two numbers preceding it (1, 1, 2, 3, 5, 8, 13, 21, 34, ...)

Float: a number which includes numbers after the decimal place

fps: frames per second

Frequency: how often something is repeated, such as the pitch of a sound

Golden mean section or ratio: depicted by the Greek letter Phi, defined by the division of a line into two parts where the longer part divided by the smaller part is equal to their sum divided by the longer part = 1.618

GUI: graphical user interface

HDD: hard-disk drive

Hertz: unit for measuring sound frequency

High-level languages: computer languages that can be read like a natural spoken language

Holographic displays: when laser light is diffracted by a specially coded film, a three-dimensional object may appear to hover in space unsupported

Hue: how much of a primary colour is present (as green is to blue and yellow)

IDE: Integrated development environment

Instance: refers to a single instance of a variable

Integer: a whole number with no decimal place

LED: light-emitting diode

Linguistics: referring to the acquisition or form of a language (European, English, Asian, or even computer)

Low-level languages: computer languages that are written so only the machine can understand

Mesh or patch: series of interconnected polygons forming a surface in two or three dimensions

Morph: to seamlessly change from one shape to another

Multisensory: using more than one of our five senses (touch, taste, smell, sight or hearing)

Neural network: a computer model for how the human brain works

Neuron: basic unit of the nervous system, nerve cell that carries electrical impulses.

Normal: direction of a vector at ninety degrees to a line or face created by a three-sided polygon

Null: returns no value

OCR: optical character recognition

Parallel: executes multiple events at the same time using different threads

Parse: to change one type of object to another (vector to raster)

Pedagogical: the act or process of teaching or learning new things

Polygon: any multi-sided object with n sides (smallest is a triangle)

Prime numbers: numbers that are not divisible by any other number except 1 (2, 3, 5, 7, 11, 13, 17, 19, 23, ...)

RAM: random access memory

Raster: a line formed by array pixels in a row

Render: display the bitmap values after calculation on a computer monitor

RGB: red, green, blue, colour gamut used in illuminated displays

Saturation: purity of a colour

Semantics: the meaning of words or phrases

Serial or sequential: executes one event at a time using a single thread

Shader: algorithm used to generate shading on a two- or three-dimensional object in a graphics program during rendering to the screen

Spline: mathematical expression for a line with at least one control point and two ends which can be distorted to form a curve

SQL: Structured Query Language

SSD: solid-state drive

Static memory: permanent memory (HDD or SSD)

String: text that can also be numbers but treated as text rather than numerically

Syntax: needed to make well formed sentences, especially for coding

Thread: the computer processing of a program using at least one core of the CPU

Timbre: difference in sounds from different types of instruments

Tint: how dark or light a colour is (as pink is to red)

Transform matrix: multiplication of values in a multi-dimensional array

Transform: to move, rotate or scale an object in two or three dimensions

Variable: a number, character, or other value that can be used over and change its value

Vector: a line with length and direction in two- or three-dimensional space

Virtual reality: fully simulated environment which one can navigate using glasses or goggles as if it were a real space

VPU: video-processing unit

Index

Milton Keynes UK
Ingram Content Group UK Ltd.
UKHW022038141024
449569UK00014B/644